Introduction to Meteorology

Introduction to
Meteorology

Edited by
Brady Walton

www.larsen-keller.com

Introduction to Meteorology
Edited by Brady Walton
ISBN: 978-1-63549-078-7 (Hardback)

© 2017 Larsen & Keller

 Larsen & Keller

Published by Larsen and Keller Education,
5 Penn Plaza,
19th Floor,
New York, NY 10001, USA

Cataloging-in-Publication Data

Introduction to meteorology / edited by Brady Walton.
      p. cm.
Includes bibliographical references and index.
ISBN 978-1-63549-078-7
1. Meteorology. 2. Atmosphere. 3. Climatology. I. Walton, Brady.
QC861.3 .I68 2017
551.5--dc23

For more information regarding Larsen and Keller Education and its products, please visit the publisher's website www.larsen-keller.com

Table of Contents

Preface   **VII**

Chapter 1 **Introduction to Meteorology**   **1**

Chapter 2 **Atmosphere: An Overview**   **14**
  • Atmosphere of Earth   14
  • Atmospheric Chemistry   52
  • Ozone Layer   56

Chapter 3 **Key Concepts of Meteorology**   **64**
  • Climate   64
  • Weather   72
  • Air Mass   79
  • Weather Front   81
  • Atmospheric Pressure   87
  • Dew Point   92
  • Precipitation   97

Chapter 4 **Understanding Important Meteorological Phenomenon**   **113**
  • Cloud   113
  • Rain   142
  • Cyclone   159
  • Cyclogenesis   169
  • Storm   173

Chapter 5 **An Integrated Study of Weather Forecasting**   **187**
  • Weather Forecasting   187
  • Weather Map   198
  • Surface Weather Analysis   204

Chapter 6 **Evolution of Meteorology**   **214**
  • Timeline of Meteorology   214

Permissions

Index

Preface

As a part of atmospheric sciences, meteorology refers to the scientific study of the atmosphere. It aims to measure and forecast atmospheric pressure and related phenomena through technology as well as the laws of physics. This book is designed specifically for students. It contains topics that are introduced to provide the readers with the basic concepts of the subject. The various sub-fields of meteorology along with technological progress that have future implications are glanced at in it. Different approaches, evaluations and methodologies have been included in this text. It will serve as a valuable source of reference for those interested in this field.

A detailed account of the significant topics covered in this book is provided below:

Chapter 1- Meteorology is the multidisciplinary study of atmospheric phenomena. It is used to explain several observable weather phenomena such as the temperature, moisture, precipitation etc. This area is very significant for the regulation of day-to-day activities. This chapter introduces the topic of meteorology while also giving a gist of the various sub-disciplines of the field.

Chapter 2- The earth's atmosphere is a complex layer of gases and water vapor. It forms a shield that filters UV radiation and is responsible for the continuation of life on Earth by retaining heat on the surface of the Earth. The atmosphere is divided into five layers based on altitude and gaseous composition- exosphere, thermosphere, mesosphere, stratosphere and troposphere. This chapter describes each of these layers and delves deep in to their discerning features. A section of the chapter deals with the ozone layer and atmospheric chemistry.

Chapter 3- This chapter provides the reader with the key concepts of meteorology like climate, weather, precipitation, air mass, atmospheric pressure, weather front, dew point and others. This helps the reader form a better understanding of these variables and their interaction. The chapter strategically encompasses and incorporates the major components and key concepts of meteorology, providing a complete understanding.

Chapter 4- Most atmospheric phenomenon are a result of changes in temperature, moisture boundary and moisture instability. Cloud formation, rain, storm and cyclones are some of the phenomenon explored in this chapter. There is a section dedicated to cyclogenesis to help the reader form a deeper understanding of the formation and/or strengthening of cyclonic circulation.

Chapter 5- This chapter focuses on the science of weather forecasting which helps predict the state of the atmosphere for a given region. It explains how weather forecasts are made and the tools, techniques and technology used for it. Listed here are topics like weather forecasting and weather map. There is a section devoted to surface weather analysis as well.

Chapter 6- This chapter chronicles the biggest scientific and technological advancements in the field of meteorology. It includes topics like observational meteorology, weather forecasting, climatology, atmospheric chemistry and atmospheric physics. It provides the reader with a timeline of meteorology spanning the ages till the present day.

I would like to make a special mention of my publisher who considered me worthy of this opportunity and also supported me throughout the process. I would also like to thank the editing team at the back-end who extended their help whenever required.

Editor

Introduction to Meteorology

Meteorology is the multidisciplinary study of atmospheric phenomena. It is used to explain several observable weather phenomena such as the temperature, moisture, precipitation etc. This area is very significant for the regulation of day-to-day activities. This chapter introduces the topic of meteorology while also giving a gist of the various sub-disciplines of the field.

Meteorology is the interdisciplinary scientific study of the atmosphere. The study of meteorology dates back millennia, though significant progress in meteorology did not occur until the 18th century. The 19th century saw modest progress in the field after weather observation networks were formed across broad regions. Prior attempts at prediction of weather depended on historical data. It wasn't until after the elucidation of the laws of physics and, more particularly, the development of the computer, allowing for the automated solution of the great many equations that model the weather, in the latter half of the 20th century that significant breakthroughs in weather forecasting were achieved.

Meteorological phenomena are observable weather events that are explained by the science of meteorology. Meteorological phenomena are described and quantified by the variables of Earth's atmosphere: temperature, air pressure, water vapor, mass flow, and the variations and interactions of those variables, and how they change over time. Different spatial scales are used to describe and predict weather on local, regional, and global levels.

Meteorology, climatology, atmospheric physics, and atmospheric chemistry are sub-disciplines of the atmospheric sciences. Meteorology and hydrology compose the interdisciplinary field of hydrometeorology. The interactions between Earth's atmosphere and its oceans are part of a coupled ocean-atmosphere system. Meteorology has application in many diverse fields such as the military, energy production, transport, agriculture, and construction.

History

Parhelion (sundog) at Savoie

The beginnings of meteorology can be traced back to ancient India, as the Upanishads contain serious discussion about the processes of cloud formation and rain and the seasonal cycles caused by the movement of Earth around the sun. Varāhamihira's classical work *Brihatsamhita*, written about 500 AD, provides clear evidence that a deep knowledge of atmospheric processes existed even in those times.

In 350 BC, Aristotle wrote *Meteorology*. Aristotle is considered the founder of meteorology. One of the most impressive achievements described in the *Meteorology* is the description of what is now known as the hydrologic cycle.

The book De Mundo (composed before 250 BC or between 350 and 200 BC) noted

> If the flashing body is set on fire and rushes violently to the Earth it is called a thunderbolt; if it be only half of fire, but violent also and massive, it is called a *meteor*; if it is entirely free from fire, it is called a smoking bolt. They are all called 'swooping bolts', because they swoop down upon the Earth. Lightning is sometimes smoky, and is then called 'smoulder- ing lightning"; sometimes it darts quickly along, and is then said to be *vivid*. At other times, it travels in crooked lines, and is called *forked lightning*. When it swoops down upon some object it is called 'swooping lightning'.

The Greek scientist Theophrastus compiled a book on weather forecasting, called the *Book of Signs*. The work of Theophrastus remained a dominant influence in the study of weather and in weather forecasting for nearly 2,000 years. In 25 AD, Pomponius Mela, a geographer for the Ro- man Empire, formalized the climatic zone system. According to Toufic Fahd, around the 9th cen- tury, Al-Dinawari wrote the *Kitab al-Nabat* (*Book of Plants*), in which he deals with the applica- tion of meteorology to agriculture during the Muslim Agricultural Revolution. He describes the meteorological character of the sky, the planets and constellations, the sun and moon, the lunar phases indicating seasons and rain, the *anwa* (heavenly bodies of rain), and atmospheric phe- nomena such as winds, thunder, lightning, snow, floods, valleys, rivers, lakes.Research of Visual Atmospheric Phenomena

Twilight at Baker Beach

Ptolemy wrote on the atmospheric refraction of light in the context of astronomical observations. In 1021, Alhazen showed that atmospheric refraction is also responsible for twilight; he estimated that twilight begins when the sun is 19 degrees below the horizon, and also used a geometric determination based on this to estimate the maximum possible height of the Earth's atmosphere as 52,000 *passuum* (about 49 miles, or 79 km).

St. Albert the Great was the first to propose that each drop of falling rain had the form of a small sphere, and that this form meant that the rainbow was produced by light interacting with each raindrop. Roger Bacon was the first to calculate the angular size of the rainbow. He stated that a rainbow summit can not appear higher than 42 degrees above the horizon. In the late 13th century and early 14th century, Kamāl al-Dīn al-Fārisī and Theodoric of Freiberg were the first to give the correct explanations for the primary rainbow phenomenon. Theoderic went further and also explained the secondary rainbow. In 1716, Edmund Halley suggested that aurorae are caused by "magnetic effluvia" moving along the Earth's magnetic field lines.

Instruments and Classification Scales

THE ROBINSON ANEMOMETER.

A hemispherical cup anemometer

In 1441, King Sejong's son, Prince Munjong, invented the first standardized rain gauge. These were sent throughout the Joseon Dynasty of Korea as an official tool to assess land taxes based upon a farmer's potential harvest. In 1450, Leone Battista Alberti developed a swinging-plate anemometer, and was known as the first *anemometer*. In 1607, Galileo Galilei constructed a thermoscope. In 1611, Johannes Kepler wrote the first scientific treatise on snow crystals: "Strena Seu de Nive Sexangula (A New Year's Gift of Hexagonal Snow)". In 1643, Evangelista Torricelli invented the mercury barometer. In 1662, Sir Christopher Wren invented the mechanical, self-emptying, tipping bucket rain gauge. In 1714, Gabriel Fahrenheit created a reliable scale for measuring temperature with a mercury-type thermometer. In 1742, Anders Celsius, a Swedish astronomer, proposed the "centigrade" temperature scale, the predecessor of the current Celsius scale. In 1783, the first hair hygrometer was demonstrated by Horace-Bénédict de Saussure. In 1802–1803, Luke Howard wrote *On the Modification of Clouds*, in which he assigns cloud types Latin names. In 1806, Francis Beaufort introduced his system for classifying wind speeds. Near the end of the 19th century the first cloud atlases were published, including the *International Cloud Atlas*, which has remained in print ever since. The April 1960 launch of the first successful weather satellite, TIROS-1, marked the beginning of the age where weather information became available globally.

Atmospheric Composition Research

In 1648, Blaise Pascal rediscovered that atmospheric pressure decreases with height, and deduced that there is a vacuum above the atmosphere. In 1738, Daniel Bernoulli published *Hydrodynamics*, initiating the Kinetic theory of gases and established the basic laws for the theory of gases. In 1761, Joseph Black discovered that ice absorbs heat without changing its temperature when melting. In 1772, Black's student Daniel Rutherford discovered nitrogen, which he called *phlogisticated air*, and together they developed the phlogiston theory. In 1777, Antoine Lavoisier discovered oxygen and developed an explanation for combustion. In 1783, in Lavoisier's essay "Reflexions sur le phlogistique", he deprecates the phlogiston theory and proposes a caloric theory. In 1804, Sir John Leslie observed that a matte black surface radiates heat more effectively than a polished surface, suggesting the importance of black body radiation. In 1808, John Dalton defended caloric theory in *A New System of Chemistry* and described how it combines with matter, especially gases; he proposed that the heat capacity of gases varies inversely with atomic weight. In 1824, Sadi Carnot analyzed the efficiency of steam engines using caloric theory; he developed the notion of a reversible process and, in postulating that no such thing exists in nature, laid the foundation for the second law of thermodynamics.

Research into Cyclones and Air Flow

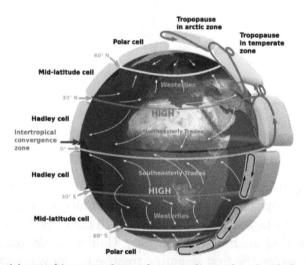

General Circulation of the Earth's Atmosphere: The westerlies and trade winds are part of the Earth's atmospheric circulation

In 1494, Christopher Columbus experienced a tropical cyclone, which led to the first written European account of a hurricane. In 1686, Edmund Halley presented a systematic study of the trade winds and monsoons and identified solar heating as the cause of atmospheric motions. In 1735, an *ideal* explanation of global circulation through study of the trade winds was written by George Hadley. In 1743, when Benjamin Franklin was prevented from seeing a lunar eclipse by a hurricane, he decided that cyclones move in a contrary manner to the winds at their periphery. Understanding the kinematics of how exactly the rotation of the Earth affects airflow was partial at first. Gaspard-Gustave Coriolis published a paper in 1835 on the energy yield of machines with rotating parts, such as waterwheels. In 1856, William Ferrel proposed the existence of a circulation cell in the mid-latitudes, and the air within deflected by the Coriolis force resulting in the prevailing westerly winds. Late in the 19th century, the motion of air masses along isobars was understood to be the result of the large-scale interaction of the pressure gradient force and the deflecting force. By 1912, this deflecting force was named the Coriolis effect. Just after World War I, a group of meteorologists in Norway led by Vilhelm Bjerknes developed the Norwegian cyclone model that explains the generation, intensification and ultimate decay (the life cycle) of mid-latitude cyclones, and introduced the idea of fronts, that is, sharply defined boundaries between air masses. The group included Carl-Gustaf Rossby (who was the first to explain the large scale atmospheric flow in terms of fluid dynamics), Tor Bergeron (who first determined how rain forms) and Jacob Bjerknes.

Observation Networks and Weather Forecasting

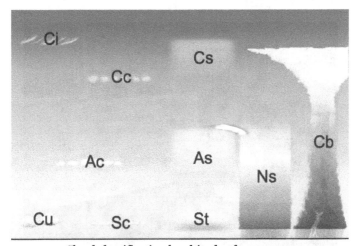

Cloud classification by altitude of occurrence

In 1654, Ferdinando II de Medici established the first *weather observing* network, that consisted of meteorological stations in Florence, Cutigliano, Vallombrosa, Bologna, Parma, Milan, Innsbruck, Osnabrück, Paris and Warsaw. The collected data were sent to Florence at regular time intervals. In 1832, an electromagnetic telegraph was created by Baron Schilling. The arrival of the electrical telegraph in 1837 afforded, for the first time, a practical method for quickly gathering surface weather observations from a wide area. This data could be used to produce maps of the state of the atmosphere for a region near the Earth's surface and to study how these states evolved through time. To make frequent weather forecasts based on these data required a reliable network of observations, but it was not until 1849 that the Smithsonian Institution began to establish an observation network across the United States under the leadership of Joseph Henry. Similar observation networks were established in Europe at this time. The Reverend William Clement Ley was key in understanding of cirrus clouds and early understandings of Jet Streams. Later after this Charles Kenneth Mackinnon Douglas known as 'CKM' Douglas read Ley's papers after his death and carried on the early study of weather systems. Nineteenth century researchers in meteorology were drawn from military or medical backgrounds, rather than trained as dedicated scientists. In 1854, the United Kingdom government appointed Robert FitzRoy to the new office of *Meteorological Statist to the Board of Trade* with the task of gathering weather observations at sea. FitzRoy's office became the United Kingdom Meteorological Office in 1854, the first national meteorological service in the world. The first daily weather forecasts made by FitzRoy's Office were published in *The Times* newspaper in 1860. The following year a system was introduced of hoisting storm warning cones at principal ports when a gale was expected.

Over the next 50 years many countries established national meteorological services. The India Meteorological Department (1875) was established to follow tropical cyclone and monsoon. The Finnish Meteorological Central Office (1881) was formed from part of Magnetic Observatory of Helsinki University. Japan's Tokyo Meteorological Observatory, the forerunner of the Japan Meteorological Agency, began constructing surface weather maps in 1883. The United States Weather Bureau (1890) was established under the United States Department of Agriculture. The Australian Bureau of Meteorology (1906) was established by a Meteorology Act to unify existing state meteorological services.

Numerical Weather Prediction

A meteorologist at the console of the IBM 7090 in the Joint Numerical Weather Prediction Unit. c. 1965

In 1904, Norwegian scientist Vilhelm Bjerknes first argued in his paper *Weather Forecasting as a Problem in Mechanics and Physics* that it should be possible to forecast weather from calculations based upon natural laws.

It was not until later in the 20th century that advances in the understanding of atmospheric physics led to the foundation of modern numerical weather prediction. In 1922, Lewis Fry Richardson published "Weather Prediction By Numerical Process", after finding notes and derivations he worked on as an ambulance driver in World War I. He described how small terms in the prognostic fluid dynamics equations that govern atmospheric flow could be neglected, and a numerical calculation scheme that could be devised to allow predictions. Richardson envisioned a large auditorium of thousands of people performing the calculations. However, the sheer number of calculations required was too large to complete without electronic computers, and the size of the grid and time steps used in the calculations led to unrealistic results. Though numerical analysis later found that this was due to numerical instability.

Starting in the 1950s, numerical forecasts with computers became feasible. The first weather forecasts derived this way used barotropic (single-vertical-level) models, and could successfully predict the large-scale movement of midlatitude Rossby waves, that is, the pattern of atmospheric lows and highs. In 1959, the UK Meteorological Office received its first computer, a Ferranti Mercury.

In the 1960s, the chaotic nature of the atmosphere was first observed and mathematically described by Edward Lorenz, founding the field of chaos theory. These advances have led to the current use of ensemble forecasting in most major forecasting centers, to take into account uncertainty arising from the chaotic nature of the atmosphere. Mathematical models used to predict the long term weather of the Earth (Climate models), have been developed that have a resolution today that are as coarse as the older weather prediction models. These climate models are used to investigate long-term climate shifts, such as what effects might be caused by human emission of greenhouse gases.

Meteorologists

Meteorologists are scientists who study meteorology. The American Meteorological Society published and continually updates an authoritative electronic *Meteorology Glossary*. Meteorologists work in government agencies, private consulting and research services, industrial enterprises, utilities, radio and television stations, and in education. In the United States, meteorologists held about 9,400 jobs in 2009.

Meteorologists are best known by the public for weather forecasting. Some radio and television weather forecasters are professional meteorologists, while others are reporters (weather specialist, weatherman, etc.) with no formal meteorological training. The American Meteorological Society and National Weather Association issue "Seals of Approval" to weather broadcasters who meet certain requirements.

Equipment

Each science has its own unique sets of laboratory equipment. In the atmosphere, there are many things or qualities of the atmosphere that can be measured. Rain, which can be observed, or seen anywhere and anytime was one of the first atmospheric qualities measured historically. Also, two other accurately measured qualities are wind and humidity. Neither of these can be seen but can be felt. The devices to measure these three sprang up in the mid-15th century and were respectively the rain gauge, the anemometer, and the hygrometer. Many attempts had been made prior to the 15th century to construct adequate equipment to measure the many atmospheric variables. Many were faulty in some way or were simply not reliable. Even Aristotle noted this in some of his work; as the difficulty to measure the air.

Sets of surface measurements are important data to meteorologists. They give a snapshot of a variety of weather conditions at one single location and are usually at a weather station, a ship or a weather buoy. The measurements taken at a weather station can include any number of atmospheric observables. Usually, temperature, pressure, wind measurements, and humidity are the variables that are measured by a thermometer, barometer, anemometer, and hygrometer, respectively. Professional stations may also include air quality sensors (carbon monoxide, carbon dioxide, methane, ozone, dust, and smoke), ceilometer (cloud ceiling), falling precipitation sensor, flood sensor, lightning sensor, microphone (explosions, sonic booms, thunder), pyranometer/pyrhelio-meter/spectroradiometer (IR/Vis/UV photodiodes), rain gauge/snow gauge, scintillation counter (background radiation, fallout, radon), seismometer (earthquakes and tremors), transmissometer (visibility), and a GPS clock for data logging. Upper air data are of crucial importance for weather forecasting. The most widely used technique is launches of radiosondes. Supplementing the radiosondes a network of aircraft collection is organized by the World Meteorological Organization.

Remote sensing, as used in meteorology, is the concept of collecting data from remote weather events and subsequently producing weather information. The common types of remote sensing are Radar, Lidar, and satellites (or photogrammetry). Each collects data about the atmosphere from a remote location and, usually, stores the data where the instrument is located. Radar and Lidar are not passive because both use EM radiation to illuminate a specific portion of the atmosphere. Weather satellites along with more general-purpose Earth-observing satellites circling the earth at various altitudes have become an indispensable tool for studying a wide range of phenomena from forest fires to El Niño.

Spatial Scales

In the study of the atmosphere, meteorology can be divided into distinct areas that depend on both time and spatial scales. At one extreme of this scale is climatology. In the timescales of hours to days, meteorology separates into micro-, meso-, and synoptic scale meteorology. Respectively, the geospatial size of each of these three scales relates directly with the appropriate timescale.

Other subclassifications are used to describe the unique, local, or broad effects within those subclasses.

Typical Scales of Atmospheric Motion Systems	
Type of motion	Horizontal scale (meter)
Molecular mean free path	10^{-3}
Minute turbulent eddies	10^{-2} - 10^{-1}
Small eddies	10^{-1} - 1
Dust devils	1 - 10
Gusts	10 - 10^{2}
Tornadoes	10^{2}
Thunderclouds	10^{3}
Fronts, squall lines	10^{4} - 10^{5}
Hurricanes	10^{5}
Synoptic Cyclones	10^{6}
Planetary waves	10^{7}
Atmospheric tides	10^{7}
Mean zonal wind	10^{7}

Microscale

Microscale meteorology is the study of atmospheric phenomena on a scale of about 1 kilometre (0.62 mi) or less. Individual thunderstorms, clouds, and local turbulence caused by buildings and other obstacles (such as individual hills) are modeled on this scale.

Mesoscale

Mesoscale meteorology is the study of atmospheric phenomena that has horizontal scales ranging from 1 km to 1000 km and a vertical scale that starts at the Earth's surface and includes the atmospheric boundary layer, troposphere, tropopause, and the lower section of the stratosphere. Mesoscale timescales last from less than a day to weeks. The events typically of interest are thunderstorms, squall lines, fronts, precipitation bands in tropical and extratropical cyclones, and topographically generated weather systems such as mountain waves and sea and land breezes.

Synoptic Scale

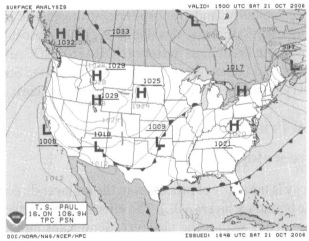

NOAA: Synoptic scale weather analysis.

Synoptic scale meteorology predicts atmosperic changes at scales up to 1000 km and 10 sec (28 days), in time and space. At the synoptic scale, the Coriolis acceleration acting on moving air masses (outside of the tropics), plays a dominant role in predictions. The phenomena typically described by synoptic meteorology include events such as extratropical cyclones, baroclinic troughs and ridges, frontal zones, and to some extent jet streams. All of these are typically given on weather maps for a specific time. The minimum horizontal scale of synoptic phenomena is limited to the spacing between surface observation stations.

Global Scale

Global scale meteorology is the study of weather patterns related to the transport of heat from the tropics to the poles. Very large scale oscillations are of importance at this scale. These oscillations have time periods typically on the order of months, such as the Madden–Julian oscillation, or years, such as the El Niño–Southern Oscillation and the Pacific decadal oscillation. Global scale meteorology pushes into the range of climatology. The traditional definition of climate is pushed into larger timescales and with the understanding of the longer time scale global oscillations, their effect on climate and weather disturbances can be included in the synoptic and mesoscale timescales predictions.

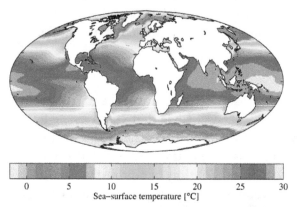

Annual mean sea surface temperatures.

Numerical Weather Prediction is a main focus in understanding air–sea interaction, tropical meteorology, atmospheric predictability, and tropospheric/stratospheric processes. The Naval Research Laboratory in Monterey, California, developed a global atmospheric model called Navy Operational Global Atmospheric Prediction System (NOGAPS). NOGAPS is run operationally at Fleet Numerical Meteorology and Oceanography Center for the United States Military. Many other global atmospheric models are run by national meteorological agencies.

Some Meteorological Principles

Boundary Layer Meteorology

Boundary layer meteorology is the study of processes in the air layer directly above Earth's surface, known as the atmospheric boundary layer (ABL). The effects of the surface – heating, cooling, and friction – cause turbulent mixing within the air layer. Significant movement of heat, matter, or momentum on time scales of less than a day are caused by turbulent motions. Boundary layer meteorology includes the study of all types of surface–atmosphere boundary, including ocean, lake, urban land and non-urban land for the study of meteorology.

Dynamic Meteorology

Dynamic meteorology generally focuses on the fluid dynamics of the atmosphere. The idea of air parcel is used to define the smallest element of the atmosphere, while ignoring the discrete molecular and chemical nature of the atmosphere. An air parcel is defined as a point in the fluid continuum of the atmosphere. The fundamental laws of fluid dynamics, thermodynamics, and motion are used to study the atmosphere. The physical quantities that characterize the state of the atmosphere are temperature, density, pressure, etc. These variables have unique values in the continuum.

Applications

Weather Forecasting

Weather forecasting is the application of science and technology to predict the state of the atmosphere at a future time and given location. Humans have attempted to predict the weather informally for millennia and formally since at least the 19th century. Weather forecasts are made by collecting quantitative data about the current state of the atmosphere and using scientific understanding of atmospheric processes to project how the atmosphere will evolve.

Forecast of surface pressures five days into the future for the north Pacific,
North America, and north Atlantic Ocean

Once an all-human endeavor based mainly upon changes in barometric pressure, current weather conditions, and sky condition, forecast models are now used to determine future conditions. Human input is still required to pick the best possible forecast model to base the forecast upon, which involves pattern recognition skills, teleconnections, knowledge of model performance, and knowledge of model biases. The chaotic nature of the atmosphere, the massive computational power required to solve the equations that describe the atmosphere, error involved in measuring the initial conditions, and an incomplete understanding of atmospheric processes mean that forecasts become less accurate as the difference in current time and the time for which the forecast is being made (the *range* of the forecast) increases. The use of ensembles and model consensus help narrow the error and pick the most likely outcome.

There are a variety of end uses to weather forecasts. Weather warnings are important forecasts because they are used to protect life and property. Forecasts based on temperature and precipitation are important to agriculture, and therefore to commodity traders within stock markets. Temperature forecasts are used by utility companies to estimate demand over coming days. On an everyday basis, people use weather forecasts to determine what to wear on a given day. Since outdoor activities are severely curtailed by heavy rain, snow and the wind chill, forecasts can be used to plan activities around these events, and to plan ahead and survive them.

Aviation Meteorology

Aviation meteorology deals with the impact of weather on air traffic management. It is important for air crews to understand the implications of weather on their flight plan as well as their aircraft, as noted by the Aeronautical Information Manual:

The effects of ice on aircraft are cumulative—thrust is reduced, drag increases, lift lessens, and weight increases. The results are an increase in stall speed and a deterioration of aircraft performance. In extreme cases, 2 to 3 inches of ice can form on the leading edge of the airfoil in less than 5 minutes. It takes but 1/2 inch of ice to reduce the lifting power of some aircraft by 50 percent and increases the frictional drag by an equal percentage.

Agricultural Meteorology

Meteorologists, soil scientists, agricultural hydrologists, and agronomists are persons concerned with studying the effects of weather and climate on plant distribution, crop yield, water-use efficiency, phenology of plant and animal development, and the energy balance of managed and natural ecosystems. Conversely, they are interested in the role of vegetation on climate and weather.

Hydrometeorology

Hydrometeorology is the branch of meteorology that deals with the hydrologic cycle, the water budget, and the rainfall statistics of storms. A hydrometeorologist prepares and issues forecasts of accumulating (quantitative) precipitation, heavy rain, heavy snow, and highlights areas with the potential for flash flooding. Typically the range of knowledge that is required overlaps with climatology, mesoscale and synoptic meteorology, and other geosciences.

The multidisciplinary nature of the branch can result in technical challenges, since tools and solutions from each of the individual disciplines involved may behave slightly differently, be optimized for different hard- and software platforms and use different data formats. There are some initiatives - such as the DRIHM project - that are trying to address this issue.

Nuclear Meteorology

Nuclear meteorology investigates the distribution of radioactive aerosols and gases in the atmosphere.

Maritime Meteorology

Maritime meteorology deals with air and wave forecasts for ships operating at sea. Organizations such as the Ocean Prediction Center, Honolulu National Weather Service forecast office, United Kingdom Met Office, and JMA prepare high seas forecasts for the world's oceans.

Military Meteorology

Military meteorology is the research and application of meteorology for military purposes. In the United States, the United States Navy's Commander, Naval Meteorology and Oceanography Command oversees meteorological efforts for the Navy and Marine Corps while the United States Air Force's Air Force Weather Agency is responsible for the Air Force and Army.

Environmental Meteorology

Environmental meteorology mainly analyzes industrial pollution dispersion physically and chemically based on meteorological parameters such as temperature, humidity, wind, and various weather conditions.

Renewable Energy

Meteorology applications in renewable energy includes basic research, "exploration", and potential mapping of wind power and solar radiation for wind and solar energy.

References

- Morelon, Régis; Rashed, Roshdi (1996). Encyclopedia of the History of Arabic Science. 3. Routledge. ISBN 0-415-12410-7.

- Jacobson, Mark Z. (June 2005). Fundamentals of Atmospheric Modeling (paperback) (2nd ed.). New York: Cambridge University Press. p. 828. ISBN 978-0-521-54865-6.

- Bluestein, H., Synoptic-Dynamic Meteorology in Midlatitudes: Principles of Kinematics and Dynamics, Vol. 1, Oxford University Press, 1992; ISBN 0-19-506267-1

- orological research in the early British straits settlements". The British Journal for the History of Science. 48 (03): 475–492. doi:10.1017/S000708741500028X. ISSN 1474-001X.

- Glickman, Todd S. (June 2009). Meteorology Glossary (electronic) (2nd ed.). Cambridge, Massachusetts: American Meteorological Society. Retrieved March 10, 2014

- Manousos, Peter (2006-07-19). "Ensemble Prediction Systems". Hydrometeorological Prediction Center. Retrieved 2010-12-31.

2

Atmosphere: An Overview

The earth's atmosphere is a complex layer of gases and water vapor. It forms a shield that filters UV radiation and is responsible for the continuation of life on Earth by retaining heat on the surface of the Earth. The atmosphere is divided into five layers based on altitude and gaseous composition-exosphere, thermosphere, mesosphere, stratosphere and troposphere. This chapter describes each of these layers and delves deep in to their discerning features. A section of the chapter deals with the ozone layer and atmospheric chemistry.

Atmosphere of Earth

The atmosphere of Earth is the layer of gases, commonly known as air, that surrounds the planet Earth and is retained by Earth's gravity. The atmosphere protects life on Earth by absorbing ultra-violet solar radiation, warming the surface through heat retention (greenhouse effect), and reducing temperature extremes between day and night (the diurnal temperature variation).

Blue light is scattered more than other wavelengths by the gases in the atmosphere, giving Earth a blue halo when seen from space onboard *ISS* at a height of 402–424 km

By volume, dry air contains 78.09% nitrogen, 20.95% oxygen, 0.93% argon, 0.039% carbon dioxide, and small amounts of other gases. Air also contains a variable amount of water vapor, on average around 1% at sea level, and 0.4% over the entire atmosphere. Air content and atmospheric pressure vary at different layers, and air suitable for use in photosynthesis by terrestrial plants and breathing of terrestrial animals is found only in Earth's troposphere and in artificial atmospheres.

The atmosphere has a mass of about 5.15×10 kg, three quarters of which is within about 11 km (6.8 mi; 36,000 ft) of the surface. The atmosphere becomes thinner and thinner with increasing altitude, with no definite boundary between the atmosphere and outer space. The Kármán line, at

100 km (62 mi), or 1.57% of Earth's radius, is often used as the border between the atmosphere and outer space. Atmospheric effects become noticeable during atmospheric reentry of spacecraft at an altitude of around 120 km (75 mi). Several layers can be distinguished in the atmosphere, based on characteristics such as temperature and composition.

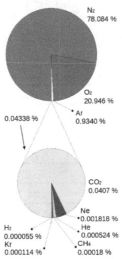

Composition of Earth's atmosphere by volume. The lower pie represents the trace gases that together compose about 0.038% of the atmosphere (0.043% with CO_2 at 2014 concentration). The numbers are from a variety of years (mainly 1987, with CO_2 and methane from 2009) and do not represent any single source.

The study of Earth's atmosphere and its processes is called atmospheric science (aerology). Early pioneers in the field include Léon Teisserenc de Bort and Richard Assmann.

Composition

Mean atmospheric water vapor

The three major constituents of air, and therefore of Earth's atmosphere, are nitrogen, oxygen, and argon. Water vapor accounts for roughly 0.25% of the atmosphere by mass. The concentration of water vapor (a greenhouse gas) varies significantly from around 10 ppm by volume in the coldest portions of the atmosphere to as much as 5% by volume in hot, humid air masses, and concentrations of other atmospheric gases are typically quoted in terms of dry air (without water vapor).The remaining gases are often referred to as trace gases,among which are the greenhouse gases, principally carbon dioxide, methane, nitrous oxide, and ozone. Filtered air includes trace amounts of many other chemical compounds. Many substances of natural origin may be present in locally and seasonally variable small amounts as aerosols in an unfiltered air sample, including dust of mineral and organic composition, pollen and spores, sea spray, and volcanic ash. Various industrial pollutants also may be present as gases or aerosols, such as chlorine (elemental or in compounds), fluorine compounds and elemental mercury vapor. Sulfur compounds such as hydrogen sulfide and sulfur dioxide (SO_2) may be derived from natural sources or from industrial air pollution.

Major constituents of dry air, by volume			
Gas		Volume[A]	
Name	Formula	in ppmv[B]	in %
Nitrogen	N_2	780,840	78.084
Oxygen	O_2	209,460	20.946
Argon	Ar	9,340	0.9340
Carbon dioxide	CO_2	397	0.0397
Neon	Ne	18.18	0.001818
Helium	He	5.24	0.000524
Methane	CH_4	1.79	0.000179
Not included in above dry atmosphere:			
Water vapor[C]	H_2O	10–50,000[D]	0.001%–5%[D]

notes:
[A] volume fraction is equal to mole fraction for ideal gas only
[B] ppmv: parts per million by volume
[C] Water vapor is about 0.25% *by mass* over full atmosphere
[D] Water vapor strongly varies locally

Structure of The Atmosphere

Principal Layers

In general, air pressure and density decrease with altitude in the atmosphere. However, temperature has a more complicated profile with altitude, and may remain relatively constant or even increase with altitude in some regions. Because the general pattern of the temperature/altitude profile is constant and measurable by means of instrumented balloon soundings, the temperature behavior provides a useful metric to distinguish atmospheric layers. In this way, Earth's atmosphere can be divided (called atmospheric stratification) into five main layers. Excluding the exosphere, Earth has four primary layers, which are the troposphere, stratosphere, mesosphere, and thermosphere. From highest to lowest, the five main layers are:

- Exosphere: 700 to 10,000 km (440 to 6,200 miles)

- Thermosphere: 80 to 700 km (50 to 440 miles)

- Mesosphere: 50 to 80 km (31 to 50 miles)

- Stratosphere: 12 to 50 km (7 to 31 miles)

- Troposphere: 0 to 12 km (0 to 7 miles)

Earth's atmosphere Lower 4 layers of the atmosphere in 3 dimensions as seen diagonally from above the exobase. Layers drawn to scale, objects within the layers are not to scale. Aurorae shown here at the bottom of the thermosphere can actually form at any altitude in this atmospheric layer

Exosphere

The exosphere is the outermost layer of Earth's atmosphere (i.e. the upper limit of the atmosphere). It extends from the exobase, which is located at the top of the thermosphere at an altitude of about 700 km above sea level, to about 10,000 km (6,200 mi; 33,000,000 ft) where it merges into the solar wind.

This layer is mainly composed of extremely low densities of hydrogen, helium and several heavier molecules including nitrogen, oxygen and carbon dioxide closer to the exobase. The atoms and molecules are so far apart that they can travel hundreds of kilometers without colliding with one another. Thus, the exosphere no longer behaves like a gas, and the particles constantly escape into space. These free-moving particles follow ballistic trajectories and may migrate in and out of the magnetosphere or the solar wind.

The exosphere is located too far above Earth for any meteorological phenomena to be possible. However, the aurora borealis and aurora australis sometimes occur in the lower part of the exosphere, where they overlap into the thermosphere. The exosphere contains most of the satellites orbiting Earth.

Thermosphere

The thermosphere is the second-highest layer of Earth's atmosphere. It extends from the mesopause (which separates it from the mesosphere) at an altitude of about 80 km (50 mi; 260,000 ft) up to the thermopause at an altitude range of 500–1000 km (310–620 mi; 1,600,000–3,300,000 ft). The height of the thermopause varies considerably due to changes in solar activity.Because the thermopause lies at the lower boundary of the exosphere, it is also referred to as the exobase. The lower part of the thermosphere, from 80 to 550 kilometres (50 to 342 mi) above Earth's surface, contains the ionosphere.

The temperature of the thermosphere gradually increases with height. Unlike the stratosphere beneath it, wherein a temperature inversion is due to the absorption of radiation by ozone, the inversion in the thermosphere occurs due to the extremely low density of its molecules. The temperature of this layer can rise as high as 1500 °C (2700 °F), though the gas molecules are so far apart that its temperature in the usual sense is not very meaningful. The air is so rarefied that an individual molecule (of oxygen, for example) travels an average of 1 kilometre (0.62 mi; 3300 ft) between collisions with other molecules.Although the thermosphere has a high proportion of molecules with high energy, it would not feel hot to a human in direct contact, because its density is too low to conduct a significant amount of energy to or from the skin.

This layer is completely cloudless and free of water vapor. However non-hydrometeorological phenomena such as the aurora borealis and aurora australis are occasionally seen in the thermosphere. The International Space Station orbits in this layer, between 350 and 420 km (220 and 260 mi).

Mesosphere

The mesosphere is the third highest layer of Earth's atmosphere, occupying the region above the stratosphere and below the thermosphere. It extends from the stratopause at an altitude of about 50 km (31 mi; 160,000 ft) to the mesopause at 80–85 km (50–53 mi; 260,000–280,000 ft) above sea level.

Temperatures drop with increasing altitude to the mesopause that marks the top of this middle layer of the atmosphere. It is the coldest place on Earth and has an average temperature around −85 °C (−120 °F; 190 K).

Just below the mesopause, the air is so cold that even the very scarce water vapor at this altitude can be sublimated into polar-mesospheric noctilucent clouds. These are the highest clouds in the atmosphere and may be visible to the naked eye if sunlight reflects off them about an hour or two after sunset or a similar length of time before sunrise. They are most readily visible when the Sun is around 4 to 16 degrees below the horizon. A type of lightning referred to as either sprites or ELVES, occasionally form far above tropospheric thunderclouds. The mesosphere is also the layer where most meteors burn up upon atmospheric entrance. It is too high above Earth to be accessible to jet-powered aircraft and balloons, and too low to permit orbital spacecraft. The mesosphere is mainly accessed by sounding rockets and rocket-powered aircraft.

Stratosphere

The stratosphere is the second-lowest layer of Earth's atmosphere. It lies above the troposphere and is separated from it by the tropopause. This layer extends from the top of the troposphere at roughly 12 km (7.5 mi; 39,000 ft) above Earth's surface to the stratopause at an altitude of about 50 to 55 km (31 to 34 mi; 164,000 to 180,000 ft).

The atmospheric pressure at the top of the stratosphere is roughly 1/1000 the pressure at sea level. It contains the ozone layer, which is the part of Earth's atmosphere that contains relatively high concentrations of that gas. The stratosphere defines a layer in which temperatures rise with increasing altitude. This rise in temperature is caused by the absorption of ultraviolet radiation (UV) radiation from the Sun by the ozone layer, which restricts turbulence and mixing. Although the temperature may be −60 °C (−76 °F; 210 K) at the tropopause, the top of the stratosphere is much warmer, and may be near 0 °C.

The stratospheric temperature profile creates very stable atmospheric conditions, so the stratosphere lacks the weather-producing air turbulence that is so prevalent in the troposphere. Consequently, the stratosphere is almost completely free of clouds and other forms of weather. However, polar stratospheric or nacreous clouds are occasionally seen in the lower part of this layer of the atmosphere where the air is coldest. This is the highest layer that can be accessed by jet-powered aircraft.

Troposphere

The troposphere is the lowest layer of Earth's atmosphere. It extends from Earth's surface to an average height of about 12 km, although this altitude actually varies from about 9 km (30,000 ft) at the poles to 17 km (56,000 ft) at the equator,with some variation due to weather. The troposphere is bounded above by the tropopause, a boundary marked in most places by a temperature inversion (i.e. a layer of relatively warm air above a colder one), and in others by a zone which is isothermal with height.

Although variations do occur, the temperature usually declines with increasing altitude in the troposphere because the troposphere is mostly heated through energy transfer from the surface. Thus, the lowest part of the troposphere (i.e. Earth's surface) is typically the warmest section of the troposphere. This promotes vertical mixing. The troposphere contains roughly 80% of the mass of Earth's atmosphere.The troposphere is denser than all its overlying atmospheric layers because a larger atmospheric weight sits on top of the troposphere and causes it to be most severely compressed. Fifty percent of the total mass of the atmosphere is located in the lower 5.6 km (18,000 ft) of the troposphere.

Nearly all atmospheric water vapor or moisture is found in the troposphere, so it is the layer where most of Earth's weather takes place. It has basically all the weather-associated cloud genus types generated by active wind circulation, although very tall cumulonimbus thunder clouds can penetrate the tropopause from below and rise into the lower part of the stratosphere. Most conventional aviation activity takes place in the troposphere, and it is the only layer that can be accessed by propeller-driven aircraft.

Space Shuttle *Endeavour* orbiting in the thermosphere. Because of the angle of the photo, it appears to straddle the stratosphere and mesosphere that actually lie more than 250 km below. The orange layer is the troposphere, which gives way to the whitish stratosphere and then the blue mesosphere.

Other Layers

Within the five principal layers that are largely determined by temperature, several secondary layers may be distinguished by other properties:

- The ozone layer is contained within the stratosphere. In this layer ozone concentrations are about 2 to 8 parts per million, which is much higher than in the lower atmosphere but still very small compared to the main components of the atmosphere. It is mainly located in the lower portion of the stratosphere from about 15–35 km (9.3–21.7 mi; 49,000–115,000 ft), though the thickness varies seasonally and geographically. About 90% of the ozone in Earth's atmosphere is contained in the stratosphere.

- The ionosphere is a region of the atmosphere that is ionized by solar radiation. It is responsible for auroras. During daytime hours, it stretches from 50 to 1,000 km (31 to 621 mi; 160,000 to 3,280,000 ft) and includes the mesosphere, thermosphere, and parts of the exosphere. However, ionization in the mesosphere largely ceases during the night, so auroras are normally seen only in the thermosphere and lower exosphere. The ionosphere forms the inner edge of the magnetosphere. It has practical importance because it influences, for example, radio propagation on Earth.

- The homosphere and heterosphere are defined by whether the atmospheric gases are well mixed. The surface-based homosphere includes the troposphere, stratosphere, mesosphere, and the lowest part of the thermosphere, where the chemical composition of the atmosphere does not depend on molecular weight because the gases are mixed by turbulence. This relatively homogeneous layer ends at the *turbopause* found at about 100 km (62 mi; 330,000 ft), which places it about 20 km (12 mi; 66,000 ft) above the mesopause.

  Above this altitude lies the heterosphere, which includes the exosphere and most of the thermosphere. Here, the chemical composition varies with altitude. This is because the distance that particles can move without colliding with one another is large compared with the size of motions that cause mixing. This allows the gases to stratify by molecular weight, with the heavier ones, such as oxygen and nitrogen, present only near the bottom of the heterosphere. The upper part of the heterosphere is composed almost completely of hydrogen, the lightest element.

- The planetary boundary layer is the part of the troposphere that is closest to Earth's surface and is directly affected by it, mainly through turbulent diffusion. During the day the planetary boundary layer usually is well-mixed, whereas at night it becomes stably stratified with weak or intermittent mixing. The depth of the planetary boundary layer ranges from as little as about 100 meters on clear, calm nights to 3000 m or more during the afternoon in dry regions.

The average temperature of the atmosphere at Earth's surface is 14 °C (57 °F; 287 K)or 15 °C (59 °F; 288 K),depending on the reference.

Physical Properties

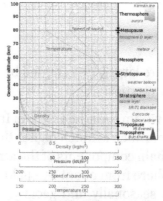

Comparison of the 1962 US Standard Atmosphere graph of geometric altitude against air density, pressure, the speed of sound and temperature with approximate altitudes of various objects.

Pressure and Thickness

The average atmospheric pressure at sea level is defined by the International Standard Atmosphere as 101325 pascals (760.00 Torr; 14.6959 psi; 760.00 mmHg). This is sometimes referred to as a unit of standard atmospheres (atm). Total atmospheric mass is 5.1480×10 kg (1.135×10 lb),about 2.5% less than would be inferred from the average sea level pressure and Earth's area of 51007.2 megahectares, this portion being displaced by Earth's mountainous terrain. Atmospheric pressure is the total weight of the air above unit area at the point where the pressure is measured. Thus air pressure varies with location and weather.

If the entire mass of the atmosphere had a uniform density from sea level, it would terminate abruptly at an altitude of 8.50 km (27,900 ft). It actually decreases exponentially with altitude, dropping by half every 5.6 km (18,000 ft) or by a factor of 1/e every 7.64 km (25,100 ft), the average scale height of the atmosphere below 70 km (43 mi; 230,000 ft). However, the atmosphere is more accurately modeled with a customized equation for each layer that takes gradients of temperature, molecular composition, solar radiation and gravity into account.

In summary, the mass of Earth's atmosphere is distributed approximately as follows:

- 50% is below 5.6 km (18,000 ft).

- 90% is below 16 km (52,000 ft).

- 99.99997% is below 100 km (62 mi; 330,000 ft), the Kármán line. By international convention, this marks the beginning of space where human travelers are considered astronauts.

By comparison, the summit of Mt. Everest is at 8,848 m (29,029 ft); commercial airliners typically cruise between 10 km (33,000 ft) and 13 km (43,000 ft) where the thinner air improves fuel economy; weather balloons reach 30.4 km (100,000 ft) and above; and the highest X-15 flight in 1963 reached 108.0 km (354,300 ft).

Even above the Kármán line, significant atmospheric effects such as auroras still occur. Meteors begin to glow in this region, though the larger ones may not burn up until they penetrate more deeply. The various layers of Earth's ionosphere, important to HF radio propagation, begin below 100 km and extend beyond 500 km. By comparison, the International Space Station and Space Shuttle typically orbit at 350–400 km, within the F-layer of the ionosphere where they encounter enough atmospheric drag to require reboosts every few months. Depending on solar activity, satellites can experience noticeable atmospheric drag at altitudes as high as 700–800 km.

Temperature and Speed of Sound

The division of the atmosphere into layers mostly by reference to temperature is discussed above. Temperature decreases with altitude starting at sea level, but variations in this trend begin above 11 km, where the temperature stabilizes through a large vertical distance through the rest of the troposphere. In the stratosphere, starting above about 20 km, the temperature increases with height, due to heating within the ozone layer caused by capture of significant ultraviolet radiation from the Sun by the dioxygen and ozone gas in this region. Still another region of increasing temperature with altitude occurs at very high altitudes, in the aptly-named thermosphere above 90 km.

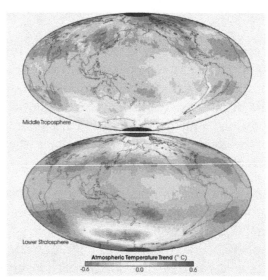

These images show temperature trends in two thick layers of the atmosphere as measured by a series
of satellite-based instruments between January 1979 and December 2005. The measurements were taken by
Microwave Sounding Units and Advanced Microwave Sounding Units flying on a series of National Oceanic and
Atmospheric Administration (NOAA) weather satellites. The instruments record microwaves emitted from oxygen
molecules in the atmosphere. Source:

Because in an ideal gas of constant composition the speed of sound depends only on temperature
and not on the gas pressure or density, the speed of sound in the atmosphere with altitude takes on
the form of the complicated temperature profile, and does not mirror altitudinal changes in density or pressure.

Density and Mass

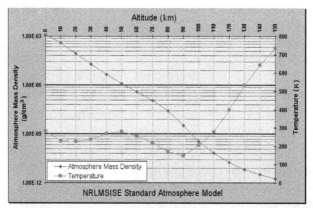

Temperature and mass density against altitude from the NRLMSISE-00 standard atmosphere model (the eight dotted
lines in each "decade" are at the eight cubes 8, 27, 64, ..., 729)

The density of air at sea level is about 1.2 kg/m (1.2 g/L, 0.0012 g/cm). Density is not measured
directly but is calculated from measurements of temperature, pressure and humidity using the
equation of state for air (a form of the ideal gas law). Atmospheric density decreases as the altitude
increases. This variation can be approximately modeled using the barometric formula. More sophisticated models are used to predict orbital decay of satellites.

The average mass of the atmosphere is about 5 quadrillion (5×10) tonnes or 1/1,200,000 the mass of Earth. According to the American National Center for Atmospheric Research, "The total mean mass of the atmosphere is 5.1480×10 kg with an annual range due to water vapor of 1.2 or 1.5×10 kg, depending on whether surface pressure or water vapor data are used; somewhat smaller than the previous estimate. The mean mass of water vapor is estimated as 1.27×10 kg and the dry air mass as 5.1352 ±0.0003×10 kg."

Optical Properties

Solar radiation (or sunlight) is the energy Earth receives from the Sun. Earth also emits radiation back into space, but at longer wavelengths that we cannot see. Part of the incoming and emitted radiation is absorbed or reflected by the atmosphere.

Scattering

When light passes through Earth's atmosphere, photons interact with it through *scattering*. If the light does not interact with the atmosphere, it is called *direct radiation* and is what you see if you were to look directly at the Sun. *Indirect radiation* is light that has been scattered in the atmosphere. For example, on an overcast day when you cannot see your shadow there is no direct radiation reaching you, it has all been scattered. As another example, due to a phenomenon called Rayleigh scattering, shorter (blue) wavelengths scatter more easily than longer (red) wavelengths. This is why the sky looks blue; you are seeing scattered blue light. This is also why sunsets are red. Because the Sun is close to the horizon, the Sun's rays pass through more atmosphere than normal to reach your eye. Much of the blue light has been scattered out, leaving the red light in a sunset.

Absorption

Different molecules absorb different wavelengths of radiation. For example, O_2 and O_3 absorb almost all wavelengths shorter than 300 nanometers. Water (H_2O) absorbs many wavelengths above 700 nm. When a molecule absorbs a photon, it increases the energy of the molecule. This heats the atmosphere, but the atmosphere also cools by emitting radiation, as discussed below.

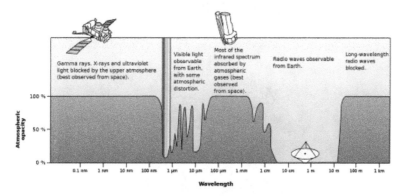

Rough plot of Earth's atmospheric transmittance (or opacity) to various wavelengths of electromagnetic radiation, including visible light.

The combined absorption spectra of the gases in the atmosphere leave "windows" of low opacity, allowing the transmission of only certain bands of light. The optical window runs from around 300 nm (ultraviolet-C) up into the range humans can see, the visible spectrum (commonly called light), at roughly 400–700 nm and continues to the infrared to around 1100 nm. There are also infrared and radio windows that transmit some infrared and radio waves at longer wavelengths. For example, the radio window runs from about one centimeter to about eleven-meter waves.

Emission

Emission is the opposite of absorption, it is when an object emits radiation. Objects tend to emit amounts and wavelengths of radiation depending on their "black body" emission curves, therefore hotter objects tend to emit more radiation, with shorter wavelengths. Colder objects emit less radiation, with longer wavelengths. For example, the Sun is approximately 6,000 K (5,730 °C; 10,340 °F), its radiation peaks near 500 nm, and is visible to the human eye. Earth is approximately 290 K (17 °C; 62 °F), so its radiation peaks near 10,000 nm, and is much too long to be visible to humans.

Because of its temperature, the atmosphere emits infrared radiation. For example, on clear nights Earth's surface cools down faster than on cloudy nights. This is because clouds (H_2O) are strong absorbers and emitters of infrared radiation. This is also why it becomes colder at night at higher elevations.

The greenhouse effect is directly related to this absorption and emission effect. Some gases in the atmosphere absorb and emit infrared radiation, but do not interact with sunlight in the visible spectrum. Common examples of these are CO_2 and H_2O.

Refractive Index

The refractive index of air is close to, but just greater than 1. Systematic variations in refractive index can lead to the bending of light rays over long optical paths. One example is that, under some circumstances, observers onboard ships can see other vessels just over the horizon because light is refracted in the same direction as the curvature of Earth's surface.

The refractive index of air depends on temperature, giving rise to refraction effects when the temperature gradient is large. An example of such effects is the mirage.

Circulation

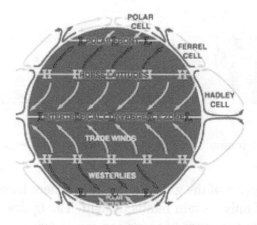

An idealised view of three large circulation cells.

Atmospheric circulation is the large-scale movement of air through the troposphere, and the means (with ocean circulation) by which heat is distributed around Earth. The large-scale structure of the atmospheric circulation varies from year to year, but the basic structure remains fairly constant because it is determined by Earth's rotation rate and the difference in solar radiation between the equator and poles.

Evolution of Earth's Atmosphere

Earliest Atmosphere

The first atmosphere would have consisted of gases in the solar nebula, primarily hydrogen. In addition, there would probably have been simple hydrides such as those now found in the gas giants (Jupiter and Saturn), notably water vapor, methane and ammonia. As the solar nebula dissipated, these gases would have escaped, partly driven off by the solar wind.

Second Atmosphere

Outgassing from volcanism, supplemented by gases produced during the late heavy bombardment of Earth by huge asteroids, produced the next atmosphere, consisting largely of nitrogen plus carbon dioxide and inert gases.A major part of carbon-dioxide emissions soon dissolved in water and built up carbonate sediments.

Researchers have found water-related sediments dating from as early as 3.8 billion years ago.About 3.4 billion years ago, nitrogen formed the major part of the then stable "second atmosphere". An influence of life has to be taken into account rather soon in the history of the atmosphere, because hints of early life-forms appear as early as 3.5 billion years ago.How Earth at that time maintained a climate warm enough for liquid water and life, if the early Sun put out 30% lower solar radiance than today, is a puzzle known as the "faint young Sun paradox".

The geological record however shows a continually relatively warm surface during the complete early temperature record of Earth - with the exception of one cold glacial phase about 2.4 billion years ago. In the late Archean eon an oxygen-containing atmosphere began to develop, apparently produced by photosynthesizing cyanobacteria, which have been found as stromatolite fossils from 2.7 billion years ago. The early basic carbon isotopy (isotope ratio proportions) very much approximates current conditions, suggesting that the fundamental features of the carbon cycle became established as early as 4 billion years ago.

Ancient sediments in the Gabon dating from between about 2,150 and 2,080 million years ago provide a record of Earth's dynamic oxygenation evolution. These fluctuations in oxygenation were likely driven by the Lomagundi carbon isotope excursion.

Third Atmosphere

The constant re-arrangement of continents by plate tectonics influences the long-term evolution of the atmosphere by transferring carbon dioxide to and from large continental carbonate stores. Free oxygen did not exist in the atmosphere until about 2.4 billion years ago during the Great Oxygenation Event and its appearance is indicated by the end of the banded iron formations. Before this time, any oxygen produced by photosynthesis was consumed by oxidation of reduced materials, notably iron. Molecules of free oxygen did not start to accumulate in the atmosphere until the rate of production of oxygen began to exceed the availability of reducing materials. This point signifies a shift from a reducing atmosphere to an oxidizing atmosphere. O_2 showed major variations until reaching a steady state of more than 15% by the end of the Precambrian.The following time span from 541 million years ago to the present day is the Phanerozoic eon, during the earliest period of which, the Cambrian, oxygen-requiring metazoan life forms began to appear.

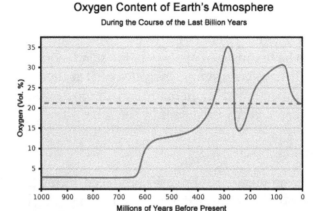

Oxygen content of the atmosphere over the last billion years. This diagram in more detail

The amount of oxygen in the atmosphere has fluctuated over the last 600 million years, reaching a peak of about 30% around 280 million years ago, significantly higher than today's 21%. Two main processes govern changes in the atmosphere: Plants use carbon dioxide from the atmosphere, releasing oxygen. Breakdown of pyrite and volcanic eruptions release sulfur into the atmosphere, which oxidizes and hence reduces the amount of oxygen in the atmosphere. However, volcanic eruptions also release carbon dioxide, which plants can convert to oxygen. The exact cause of the variation of the amount of oxygen in the atmosphere is not known. Periods with much oxygen in the atmosphere are associated with rapid development of animals. Today's atmosphere contains 21% oxygen, which is high enough for this rapid development of animals.

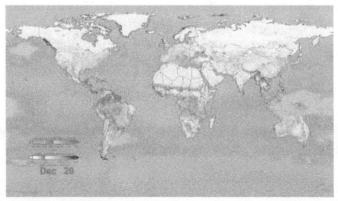

This animation shows the buildup of tropospheric CO_2 in the Northern Hemisphere with a maximum around May. The maximum in the vegetation cycle follows, occurring in the late summer. Following the peak in vegetation, the drawdown of atmospheric CO_2 due to photosynthesis is apparent, particularly over the boreal forests.

The scientific consensus is that the anthropogenic greenhouse gases currently accumulating in the atmosphere are the main cause of global warming.

Air Pollution

Air pollution is the introduction into the atmosphere of chemicals, particulate matter or biological materials that cause harm or discomfort to organisms.Stratospheric ozone depletion is caused by air pollution, chiefly from chlorofluorocarbons and other ozone-depleting substances.

Images from Space

On October 19, 2015 NASA started a website containing daily images of the full sunlit side of Earth. The images are taken from the Deep Space Climate Observatory (DSCOVR) and show Earth as it rotates during a day.

Different Layers of Atmosphere

Exosphere

Earth atmosphere diagram showing the exosphere and other layers. The layers are to scale. From Earth's surface to the top of the stratosphere (50km) is just under 1% of Earth's radius.

The exosphere is a thin, atmosphere-like volume surrounding a planet or natural satellite where molecules are gravitationally bound to that body, but where the density is too low for them to behave as a gas by colliding with each other.In the case of bodies with substantial atmospheres, such as Earth's atmosphere, the exosphere is the uppermost layer, where the atmosphere thins out and merges with interplanetary space. It is located directly above the thermosphere.

Limb view, of Earth's atmosphere. Colors roughly denote the layers of the atmosphere.

The geomagnetic storms cause beautiful displays of aurora across the atmosphere.

Several moons, such as the Moon and the Galilean satellites of Jupiter, have exospheres without a denser atmosphere underneath,referred to as a surface boundary exosphere.Here, molecules are ejected on elliptic trajectories until they collide with the surface. Smaller bodies such as asteroids, in which the molecules emitted from the surface escape to space, are not considered to have exospheres.

Earth's Exosphere

The most common molecules within Earth's exosphere are those of the lightest atmospheric gasses. Hydrogen is present throughout the exosphere, with some helium, carbon dioxide, and atomic oxygen near its base. Because it can be difficult to define the boundary between the exosphere and outer space the exosphere may be considered a part of interplanetary or outer space.

Lower Boundary

The lower boundary of the exosphere is called the *exobase*. It is also called *exopause* and *'critical altitude'* as this is the altitude where barometric conditions no longer apply. Atmospheric temperature becomes nearly a constant above this altitude.On Earth, the altitude of the exobase ranges from about 500 to 1,000 kilometres (310 to 620 mi) depending on solar activity.

The exobase can be defined in one of two ways:

If we define the exobase as the height at which upward-traveling molecules experience one collision on average, then at this position the mean free path of a molecule is equal to one pressure scale height. This is shown in the following. Consider a volume of air, with horizontal area A and height equal to the mean free path l , at pressure p and temperature T. For an ideal gas, the number of molecules contained in it is:

$$n = \frac{pAl}{RT}$$

where R is the universal gas constant. From the requirement that each molecule traveling upward undergoes on average one collision, the pressure is:

$$p = \frac{m_A ng}{A}$$

where m_A is the mean molecular mass of the gas. Solving these two equations gives:

$$l = \frac{RT}{m_A g}$$

which is the equation for the pressure scale height. As the pressure scale height is almost equal to the density scale height of the primary constituent, and because the Knudsen number is the ratio of mean free path and typical density fluctuation scale, this means that the exobase lies in the region where $Kn(h_{EB}) \simeq 1$.

The fluctuation in the height of the exobase is important because this provides atmospheric drag on satellites, eventually causing them to fall from orbit if no action is taken to maintain the orbit.

Upper Boundary of Earth

In principle, the exosphere covers distances where particles are still gravitationally bound to Earth, i.e. particles still have ballistic orbits that will take them back towards Earth. The upper boundary of the exosphere can be defined as the distance at which the influence of solar radiation pressure on atomic hydrogen exceeds that of Earth's gravitational pull. This happens at half the distance to the Moon (the average distance between Earth and the Moon is 384,400 kilometres (238,900 mi)). The exosphere, observable from space as the geocorona, is seen to extend to at least 10,000 kilometres (6,200 mi) from Earth's surface. The exosphere is a transitional zone between Earth's atmosphere and space.

Moon's Exosphere

On 17 August 2015, based on studies with the Lunar Atmosphere and Dust Environment Explorer (LADEE) spacecraft, NASA scientists reported the detection of neon in the exosphere of the moon.

Thermosphere

The thermosphere is the layer of the Earth's atmosphere directly above the mesosphere and directly below the exosphere. Within this layer of the atmosphere, ultraviolet radiation causes photoionization/photodissociation of molecules, creating ions in the ionosphere. At these high altitudes, the residual atmospheric gases sort into strata according to molecular mass. Thermospheric temperatures increase with altitude due to absorption of highly energetic solar radiation. Temperatures are highly dependent on solar activity, and can rise to 2,000°C (3,630°F). Radiation causes the atmosphere particles in this layer to become electrically charged, enabling radio waves to be refracted and thus be received beyond the horizon. In the exosphere, beginning at 500 to 1,000 kilometres (310 to 620 mi) above the Earth's surface, the atmosphere turns into space.

The highly diluted gas in this layer can reach 2,500 °C (4,530 °F) during the day. Even though the temperature is so high, one would not feel warm in the thermosphere, because it is so near vacuum that there is not enough contact with the few atoms of gas to transfer much heat. A normal thermometer might be significantly below 0 °C (32 °F), at least at night, because the energy lost by thermal radiation would exceed the energy acquired from the atmospheric gas by direct contact. In the anacoustic zone above 160 kilometres (99 mi), the density is so low that molecular interactions are too infrequent to permit the transmission of sound.

The dynamics of the thermosphere are dominated by atmospheric tides, which are driven by the very significant diurnal heating. Atmospheric waves dissipate above this level because of collisions between the neutral gas and the ionospheric plasma.

The International Space Station orbits the Earth within the middle of the thermosphere, between 330 and 435 kilometres (205 and 270 mi) (decaying by 2 km/month and raised by periodic reboosts), whereas the Gravity Field and Steady-State Ocean Circulation Explorer satellite at 260 kilometres (160 mi) utilized wingletsand an innovative ion engine to maintain a stable orientation and orbit.

Neutral gas constituents

It is convenient to separate the atmospheric regions according to the two temperature minima at about 12 km altitude (the tropopause) and at about 85 km (the mesopause) (Figure 1). The thermosphere (or the upper atmosphere) is the height region above 85 km, while the region between the tropospause and the mesopause is the middle atmosphere (stratosphere and mesosphere) where absorption of solar UV radiation generates the temperature maximum near 45 km altitude and causes the ozone layer.

The density of the Earth's atmosphere decreases nearly exponentially with altitude. The total mass of the atmosphere is $M = \rho_A H$ 1 kg/cm within a column of one square centimeter above the ground (with $\rho_A = 1.29$ kg/m the atmospheric density on the ground at z = 0 m altitude, and H 8 km the average atmospheric scale height). 80% of that mass is concentrated within the troposphere. The mass of the thermosphere above about 85 km is only 0.002% of the total mass. Therefore, no significant energetic feedback from the thermosphere to the lower atmospheric regions can be expected.

Turbulence causes the air within the lower atmospheric regions below the turbopause at about 110 km to be a mixture of gases that does not change its composition. Its mean molecular weight is 29 g/mol with molecular oxygen (O_2) and nitrogen (N_2) as the two dominant constituents. Above the turbopause, however, diffusive separation of the various constituents is significant, so that each constituent follows its own barometric height structure with a scale height inversely proportional to its molecular weight. The lighter constituents atomic oxygen (O), helium (He), and hydrogen (H) successively dominate above about 200 km altitude and vary with geographic location, time, and solar activity. The ratio N_2/O which is a measure of the electron density at the ionospheric F region is highly affected by these variations.These changes follow from the diffusion of the minor constituents through the major gas component during dynamic processes.

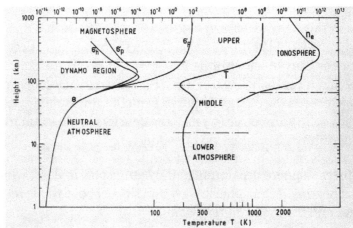

Nomenclature of atmospheric regions based on the profiles of electric conductivity (left), temperature (middle), and electron density (right)

The thermosphere contains an appreciable concentration of elemental sodium located in a 10-km thick band that occurs at the edge of the mesosphere, 80 to 100 km above Earth's surface. The sodium has an average concentration of 400,000 atoms per cubic centimeter. This band is regularly replenished by sodium sublimating from incoming meteors. Astronomers have begun utilizing this sodium band to create "guide stars" as part of the optical correction process in producing ultra-sharp ground-based observations.

Energy Input

Energy Budget

The thermospheric temperature can be determined from density observations as well as from direct satellite measurements. The temperature vs. altitude z in Fig. 1 can be simulated by the so-called Bates profile:

(1)
$$T = T_\infty - (T_\infty - T_0)\exp\{-s(z - z_0)\}$$

with T_∞ the exospheric temperature above about 400 km altitude, T_0 = 355 K, and z_0 = 120 km reference temperature and height, and s an empirical parameter depending on T_∞ and decreasing with T_∞. That formula is derived from a simple equation of heat conduction. One estimates a total heat input of q_0 0.8 to 1.6 mW/m above z_0 = 120 km altitude. In order to obtain equilibrium conditions, that heat input q_0 above z_0 is lost to the lower atmospheric regions by heat conduction.

The exospheric temperature T_∞ is a fair measurement of the solar XUV radiation. Since solar radio emission F at 10.7 cm wavelength is a good indicator of solar activity, one can apply the empirical formula for quiet magnetospheric conditions.

(2)
$$T_\infty \simeq 500 + 3.4F_0$$

with T_∞ in K, F_0 in 10^{-2} Wm Hz (the Covington index) a value of F averaged over several solar cycles. The Covington index varies typically between 70 and 250 during a solar cycle, and never drops below about 50. Thus, T_∞ varies between about 740 and 1350 K. During very quiet magnetospheric conditions, the still continuously flowing magnetospheric energy input contributes by about 250 K to the residual temperature of 500 K in eq.(2). The rest of 250 K in eq.(2) can be attributed to atmospheric waves generated within the troposphere and dissipated within the lower thermosphere.

Solar XUV Radiation

The solar X-ray and extreme ultraviolet radiation (XUV) at wavelengths < 170 nm is almost completely absorbed within the thermosphere. This radiation causes the various ionospheric layers as well as a temperature increase at these heights. While the solar visible light (380 to 780 nm) is nearly constant with a variability of not more than about 0.1% of the solar constant, the solar XUV radiation is highly variable in time and space. For instance, X-ray bursts associated with solar flares can dramatically increase their intensity over preflare levels by many orders of magnitude over a time span of tens of minutes. In the extreme ultraviolet, the Lyman α line at

121.6 nm represents an important source of ionization and dissociation at ionospheric D layer heights.During quiet periods of solar activity, it alone contains more energy than the rest of the XUV spectrum. Quasi-periodic changes of the order of 100% or greater, with periods with period of 27 days and 11 years, belong to the prominent variations of solar XUV radiation. However, irregular fluctuations over all time scales are present all the time.During low solar activity, about half of the total energy input into the thermosphere is thought to be solar XUV radiation. Evidently, that solar XUV energy input occurs only during daytime conditions, maximizing at the equator during equinox.

Solar Wind

A second source of energy input into the thermosphere is solar wind energy which is transferred to the magnetosphere by mechanisms that are not well understood. One possible way to transfer energy is via a hydrodynamic dynamo process. Solar wind particles penetrate into the polar regions of the magnetosphere where the geomagnetic field lines are essentially vertically directed. An electric field is generated, directed from dawn to dusk. Along the last closed geomagnetic field lines with their footpoints within the auroral zones, field aligned electric currents can flow into the ionospheric dynamo region where they are closed by electric Pedersen and Hall currents. Ohmic losses of the Pedersen currents heat the lower thermosphere. In addition, penetration of high energetic particles from the magnetosphere into the auroral regions enhance drastically the electric conductivity, further increasing the electric currents and thus Joule heating. During quiet magnetospheric activity, the magnetosphere contributes perhaps by a quarter to the thermosphere's energy budget.This is about 250 K of the exospheric temperature in eq.(2). During very large activity, however, this heat input can increase substantially, by a factor of four or more. That solar wind input occurs mainly in the auroral regions during both day and night.

Atmospheric Waves

Two kinds of large-scale atmospheric waves within the lower atmosphere exist: internal waves with finite vertical wavelengths which can transport wave energy upward; and external waves with infinitely large wavelengths which cannot transport wave energy.Atmospheric gravity waves and most of the atmospheric tides generated within the troposphere belong to the internal waves. Their density amplitudes increase exponentially with height, so that at the mesopause these waves become turbulent and their energy is dissipated (similar to breaking of ocean waves at the coast), thus contributing to the heating of the thermosphere by about 250 K in eq.(2). On the other hand, the fundamental diurnal tide labelled (1, −2) which is most efficiently excited by solar irradiance is an external wave and plays only a marginal role within lower and middle atmosphere. However, at thermospheric altitudes, it becomes the predominant wave. It drives the electric Sq-current within the ionospheric dynamo region between about 100 and 200 km height.

Heating, predominately by tidal waves, occurs mainly at lower and middle latitudes. The variability of this heating depends on the meteorological conditions within troposphere and middle atmosphere, and may not exceed about 50%.

Dynamics

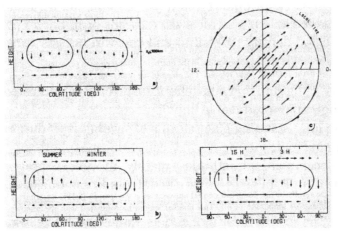

Figure 2. Schematic meridian-height cross-section of circulation of (a) symmetric wind component (P_2), (b) of antisymmetric wind component (P_1), and (d) of symmetric diurnal wind component (P_1) at 3 h and 15 h local time. Upper right pannel (c) shows the horizontal wind vectors of the diurnal component in the northern hemisphere depending on local time.

Within the thermosphere above about 150 km height, all atmospheric waves successively become external waves, and no significant vertical wave structure is visible. The atmospheric wave modes degenerate to the spherical functions P_n with m a meridional wave number and n the zonal wave number (m = 0: zonal mean flow; m = 1: diurnal tides; m = 2: semidiurnal tides; etc.). The thermosphere becomes a damped oscillator system with low-pass filter characteristics. This means that smaller-scale waves (greater numbers of (n,m)) and higher frequencies are suppressed in favor of large-scale waves and lower frequencies. If one considers very quiet magnetospheric disturbances and a constant mean exospheric temperature (averaged over the sphere), the observed temporal and spatial distribution of the exospheric temperature distribution can be described by a sum of spheric functions:

$$(3) \qquad T(\varphi,\lambda,t) = T_\infty \{1 + \Delta T_2^0 P_2^0(\varphi) + \Delta T_1^0 P_1^0(\varphi)\cos[\omega_a(t-t_a)] + \Delta T_1^1 P_1^1(\varphi)\cos(\tau - \tau_d) + ...\}$$

Here, it is φ latitude, λ longitude, and t time, ω_a the angular frequency of one year, ω_d the angular frequency of one solar day, and $\tau = \omega_d t + \lambda$ the local time. t_a = June 21 is the date of northern summer solstice, and τ_d = 15:00 is the local time of maximum diurnal temperature.

The first term in (3) on the right is the global mean of the exospheric temperature (of the order of 1000 K). The second term [with P_2 = 0.5(3 sin(φ)–1)] represents the heat surplus at lower latitudes and a corresponding heat deficit at higher latitudes (Fig. 2a). A thermal wind system develops with wind toward the poles in the upper level and wind away from the poles in the lower level. The coefficient $\Delta T_2 \approx 0.004$ is small because Joule heating in the aurora regions compensates that heat surplus even during quiet magnetospheric conditions. During disturbed conditions, however, that term becomes dominant, changing sign so that now heat surplus is transported from the poles to the equator. The third term (with P_1 = sin φ) represents heat surplus on the summer hemisphere and is responsible for the transport of excess heat from the summer into the winter hemisphere (Fig. 2b). Its relative amplitude is of the order ΔT_1 0.13. The fourth term (with $P_1(\varphi)$ = cos φ) is the dominant diurnal wave (the tidal mode (1,–2)). It is responsible for the transport of excess heat from the daytime hemisphere into the nighttime hemisphere. Its relative amplitude is ΔT_1 0.15, thus on the order of 150 K. Additional terms (e.g., semiannual, semidiurnal terms and higher order terms) must be added to eq.(3). However, they are of minor importance. Corresponding sums can be developed for density, pressure, and the various gas constituents.Thermospheric Storms.

In contrast to solar XUV radiation, magnetospheric disturbances, indicated on the ground by geomagnetic variations, show an unpredictable impulsive character, from short periodic disturbances of the order of hours to long-standing giant storms of several day's duration. The reaction of the thermosphere to a large magnetospheric storm is called thermospheric storm. Since the heat input into the thermosphere occurs at high latitudes (mainly into the auroral regions), the heat transport represented by the term P_2 in eq.(3) is reversed. In addition, due to the impulsive form of the disturbance, higher-order terms are generated which, however, possess short decay times and thus quickly disappear. The sum of these modes determines the "travel time" of the disturbance to the lower latitudes, and thus the response time of the thermosphere with respect to the magnetospheric disturbance. Important for the development of an ionospheric storm is the increase of the ratio N_2/O during a thermospheric storm at middle and higher latitude. An increase of N_2 increases the loss process of the ionospheric plasma and causes therefore a decrease of the electron density within the ionospheric F-layer (negative ionospheric storm).

Mesosphere

The mesosphere is the layer of the Earth's atmosphere that is directly above the stratosphere and directly below the mesopause. In the mesosphere, temperature decreases as the altitude increases. The upper boundary of the mesosphere is the mesopause, which can be the coldest naturally occurring place on Earth with temperatures below 143°C (−225°F;−130°K). The exact upper and lower boundaries of the mesosphere vary with latitude and with season, but the lower boundary of the mesosphere is usually located at heights of about 50 kilometres (160,000 ft; 31 mi) above the Earth's surface and the mesopause is usually at heights near 100 kilometres (62 mi), except at middle and high latitudes in summer where it descends to heights of about 85 kilometres (53 mi).

The stratosphere, mesosphere and lowest part of the thermosphere are collectively referred to as the "middle atmosphere", which spans heights from approximately 10 kilometres (33,000 ft) to 100 kilometres (62 mi). The mesopause, at an altitude of 80–90 km (50–56 mi), separates the mesosphere from the thermosphere—the second-outermost layer of the Earth's atmosphere. This is also around the same altitude as the turbopause, below which different chemical species are well mixed due to turbulent eddies. Above this level the atmosphere becomes non-uniform; the scale heights of different chemical species differ by their molecular masses.

Temperature

Within the mesosphere, temperature decreases with increasing height, due to decreasing solar heating and increasing cooling by CO_2 radiative emission. The top of the mesosphere, called the mesopause, is the coldest part of Earth's atmosphere.Temperatures in the upper mesosphere fall as low as −101°C (172 K; −150°F),varying according to latitude and season.

Dynamic Features

The main dynamic features in this region are strong zonal (East-West) winds, atmospheric tides, internal atmospheric gravity waves (commonly called "gravity waves"), and planetary waves. Most of these tides and waves start in the troposphere and lower stratosphere, and propagate to the mesosphere. In the mesosphere, gravity-wave amplitudes can become so large that the waves become unstable and dissipate. This dissipation deposits momentum into the mesosphere and largely drives global circulation.

Noctilucent clouds are located in the mesosphere. The upper mesosphere is also the region of the ionosphere known as the *D layer*. The D layer is only present during the day, when some ionization occurs with nitric oxide being ionized by Lyman series-alpha hydrogen radiation. The ionization is so weak that when night falls, and the source of ionization is removed, the free electron and ion form back into a neutral molecule. The mesosphere is also known as the "Ignorosphere" because it is poorly studied compared to the stratosphere (which can be accessed with high-altitude balloons) and the thermosphere (in which satellites can orbit).

A 5 km (3.1 mi) deep sodium layer is located between 80–105 km (50–65 mi). Made of unbound, non-ionized atoms of sodium, the sodium layer radiates weakly to contribute to the airglow. The sodium has an average concentration of 400,000 atoms per cubic centimeter. This band is regularly replenished by sodium sublimating from incoming meteors. Astronomers have begun utilizing this sodium band to create "guide stars" as part of the optical correction process in producing ultra-sharp ground-based observations.

Millions of meteors enter the Earth's atmosphere, an average of 40 tons per year.

Uncertainties

The mesosphere lies above altitude records for aircraft and balloons,and below the minimum altitude for orbital spacecraft.It has only been accessed through the use of sounding rockets.As a result, it is the least-understood part of the atmosphere. The presence of red sprites and blue jets (electrical discharges or lightning within the lower mesosphere), noctilucent clouds, and density shears within this poorly understood layer are of current scientific interest.

Ionosphere

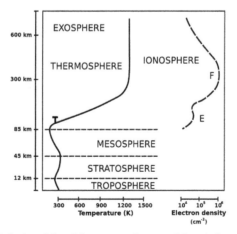

Relationship of the atmosphere and ionosphere

The ionosphere is a region of Earth's upper atmosphere, from about 60 km (37 mi) to 1,000 km (620 mi) altitude,and includes the thermosphere and parts of the mesosphere and exosphere. It is ionized by solar radiation, plays an important part in atmospheric electricity and forms the inner edge of the magnetosphere. It has practical importance because, among other functions, it influences radio propagation to distant places on the Earth.

Geophysics

The ionosphere is a shell of electrons and electrically charged atoms and molecules that surrounds the Earth, stretching from a height of about 50 km (31 mi) to more than 1,000 km (620 mi). It owes its existence primarily to ultraviolet radiation from the Sun.

The lowest part of the Earth's atmosphere, the troposphere extends from the surface to about 10 km (6.2 mi). Above 10 km (6.2 mi) is the stratosphere, followed by the mesosphere. In the stratosphere incoming solar radiation creates the ozone layer. At heights of above 80 km (50 mi), in the thermosphere, the atmosphere is so thin that free electrons can exist for short periods of time before they are captured by a nearby positive ion. The number of these free electrons is sufficient to affect radio propagation. This portion of the atmosphere is *ionized* and contains a plasma which is referred to as the ionosphere. In a plasma, the negative free electrons and the positive ions are attracted to each other by the electrostatic force, but they are too energetic to stay fixed together in an electrically neutral molecule.

Ultraviolet (UV), X-ray and shorter wavelengths of solar radiation are *ionizing,* since photons at these frequencies contain sufficient energy to dislodge an electron from a neutral gas atom or molecule upon absorption. In this process the light electron obtains a high velocity so that the temperature of the created electronic gas is much higher (of the order of thousand K) than the one of ions and neutrals. The reverse process to ionization is recombination, in which a free electron is "captured" by a positive ion. Recombination occurs spontaneously, and causes the emission of a photon carrying away the energy produced upon recombination. As gas density increases at lower altitudes, the recombination process prevails, since the gas molecules and ions are closer together. The balance between these two processes determines the quantity of ionization present.

Ionization depends primarily on the Sun and its activity. The amount of ionization in the ionosphere varies greatly with the amount of radiation received from the Sun. Thus there is a diurnal (time of day) effect and a seasonal effect. The local winter hemisphere is tipped away from the Sun, thus there is less received solar radiation. The activity of the Sun is associated with the sunspot cycle, with more radiation occurring with more sunspots. Radiation received also varies with geographical location (polar, auroral zones, mid-latitudes, and equatorial regions). There are also mechanisms that disturb the ionosphere and decrease the ionization. There are disturbances such as solar flares and the associated release of charged particles into the solar wind which reaches the Earth and interacts with its geomagnetic field.

The Ionospheric Layers

At night the F layer is the only layer of significant ionization present, while the ionization in the E and D layers is extremely low. During the day, the D and E layers become much more heavily ionized, as does the F layer, which develops an additional, weaker region of ionisation known as the F_1

layer. The F_2 layer persists by day and night and is the region mainly responsible for the refraction of radio waves.

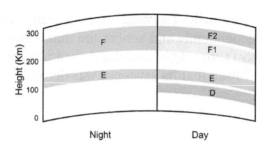

Ionospheric layers.

D Layer

The D layer is the innermost layer, 60 km (37 mi) to 90 km (56 mi) above the surface of the Earth. Ionization here is due to Lyman series-alpha hydrogen radiation at a wavelength of 121.5 nanometre (nm) ionizing nitric oxide (NO). In addition, high solar activity can generate hard X-rays (wavelength < 1 nm) that ionize N_2 and O_2. Recombination rates are high in the D layer, so there are many more neutral air molecules than ions. Medium frequency (MF) and lower high frequency (HF) radio waves are significantly reduced in strength within the D layer, as the passing radio waves cause electrons to move, which then collide with the neutral molecules, giving up their energy. The lower frequencies move the electrons farther, with a greater chance of collisions. This is the main reason for absorption of HF radio waves, particularly at 10 MHz and below, with progressively less absorption at higher frequencies. This effect peaks around noon and is reduced at night due to a decrease in the D layer's thickness; only a small part remains due to cosmic rays. A common example of the D layer in action is the disappearance of distant AM broadcast band stations in the daytime.

During solar proton events, ionization can reach unusually high levels in the D-region over high and polar latitudes. Such very rare events are known as Polar Cap Absorption (or PCA) events, because the increased ionization significantly enhances the absorption of radio signals passing through the region. In fact, absorption levels can increase by many tens of dB during intense events, which is enough to absorb most (if not all) transpolar HF radio signal transmissions. Such events typically last less than 24 to 48 hours.

E Layer

The E layer is the middle layer, 90 km (56 mi) to 150 km (93 mi) above the surface of the Earth. Ionization is due to soft X-ray (1–10 nm) and far ultraviolet (UV) solar radiation ionization of molecular oxygen (O_2). Normally, at oblique incidence, this layer can only reflect radio waves having frequencies lower than about 10 MHz and may contribute a bit to absorption on frequencies above. However, during intense Sporadic E events, the E_s layer can reflect frequencies up to 50 MHz and higher. The vertical structure of the E layer is primarily determined by the competing effects of ionization and recombination. At night the E layer weakens because the primary source of ionization is no longer present. After sunset an increase in the height of the E layer maximum increases the range to which radio waves can travel by reflection from the layer.

This region is also known as the Kennelly–Heaviside layer or simply the Heaviside layer. Its existence was predicted in 1902 independently and almost simultaneously by the American electrical engineer Arthur Edwin Kennelly (1861–1939) and the British physicist Oliver Heaviside (1850–1925). However, it was not until 1924 that its existence was detected by Edward V. Appleton and Miles Barnett.

E_s

The E_s layer (sporadic E-layer) is characterized by small, thin clouds of intense ionization, which can support reflection of radio waves, rarely up to 225 MHz. Sporadic-E events may last for just a few minutes to several hours. Sporadic E propagation makes VHF-operating radio amateurs very excited, as propagation paths that are generally unreachable can open up. There are multiple causes of sporadic-E that are still being pursued by researchers. This propagation occurs most frequently during the summer months when high signal levels may be reached. The skip distances are generally around 1,640 km (1,020 mi). Distances for one hop propagation can be anywhere from 900 km (560 mi) to 2,500 km (1,600 mi). Double-hop reception over 3,500 km (2,200 mi) is possible.

F layer

The F layer or region, also known as the Appleton-Barnett layer, extends from about 150 km (93 mi) to more than 500 km (310 mi) above the surface of Earth. It is the densest point of the ionosphere, which implies signals penetrating this layer will escape into space. At higher altitudes, the number of oxygen ions decreases and lighter ions such as hydrogen and helium become dominant; this layer is the topside ionosphere. There, extreme ultraviolet (UV, 10–100 nm) solar radiation ionizes atomic oxygen. The F layer consists of one layer at night, but during the day, a deformation often forms in the profile that is labeled F1. The F2 layer remains by day and night responsible for most skywave propagation of radio waves, facilitating high frequency (HF, or shortwave) radio communications over long distances.

From 1972 to 1975 NASA launched the AEROS and AEROS B satellites to study the F region.

Ionospheric Model

An ionospheric model is a mathematical description of the ionosphere as a function of location, altitude, day of year, phase of the sunspot cycle and geomagnetic activity. Geophysically, the state of the ionospheric plasma may be described by four parameters: *electron density, electron and ion temperature* and, since several species of ions are present, *ionic composition*. Radio propagation depends uniquely on electron density.

Models are usually expressed as computer programs. The model may be based on basic physics of the interactions of the ions and electrons with the neutral atmosphere and sunlight, or it may be a statistical description based on a large number of observations or a combination of physics and observations. One of the most widely used models is the International Reference Ionosphere (IRI), which is based on data and specifies the four parameters just mentioned. The IRI is an international project sponsored by the Committee on Space Research (COSPAR) and the International Union of Radio Science (URSI).The major data sources are the worldwide network of ionosondes, the powerful incoherent scatter radars (Jicamarca, Arecibo, Millstone Hill, Malvern, St. Santin), the ISIS and Alouette topside sounders, and in situ instruments on several satellites and rockets. IRI is updated yearly. IRI is more accurate in describing the variation of the electron density from bottom of the ionosphere to the altitude of maximum density than in describing the total electron content (TEC).

Persistent Anomalies to The Idealized Model

Ionograms allow deducing, via computation, the true shape of the different layers. Nonhomogeneous structure of the electron/ion-plasma produces rough echo traces, seen predominantly at night and at higher latitudes, and during disturbed conditions.

Winter Anomaly

At mid-latitudes, the F_2 layer daytime ion production is higher in the summer, as expected, since the Sun shines more directly on the Earth. However, there are seasonal changes in the molecular-to-atomic ratio of the neutral atmosphere that cause the summer ion loss rate to be even higher. The result is that the increase in the summertime loss overwhelms the increase in summertime production, and total F_2 ionization is actually lower in the local summer months. This effect is known as the winter anomaly. The anomaly is always present in the northern hemisphere, but is usually absent in the southern hemisphere during periods of low solar activity.

Equatorial Anomaly

Electric currents created in sunward ionosphere.

Within approximately ± 20 degrees of the *magnetic equator*, is the *equatorial anomaly*. It is the occurrence of a trough in the ionization in the F_2 layer at the equator and crests at about 17 degrees in magnetic latitude. The Earth's magnetic field lines are horizontal at the magnetic equator. Solar heating and tidal oscillations in the lower ionosphere move plasma up and across the magnetic field lines. This sets up a sheet of electric current in the E region which, with the horizontal magnetic field, forces ionization up into the F layer, concentrating at ± 20 degrees from the magnetic equator. This phenomenon is known as the *equatorial fountain*.

Equatorial Electrojet

The worldwide solar-driven wind results in the so-called Sq (solar quiet) current system in the E region of the Earth's ionosphere (ionospheric dynamo region) (100 km (62 mi) – 130 km (81 mi) altitude). Resulting from this current is an electrostatic field directed E-W (dawn-dusk) in the equatorial day side of the ionosphere. At the magnetic dip equator, where the geomagnetic field is horizontal, this electric field results in an enhanced eastward current flow within ± 3 degrees of the magnetic equator, known as the equatorial electrojet.

Ephemeral Ionospheric Perturbations

X-rays: Sudden Ionospheric Disturbances (SID)

When the Sun is active, strong solar flares can occur that will hit the sunlit side of Earth with hard X-rays. The X-rays will penetrate to the D-region, releasing electrons that will rapidly increase absorption, causing a high frequency (3–30 MHz) radio blackout. During this time very low frequency (3–30 kHz) signals will be reflected by the D layer instead of the E layer, where the increased atmospheric density will usually increase the absorption of the wave and thus dampen it. As soon as the X-rays end, the sudden ionospheric disturbance (SID) or radio black-out ends as the electrons in the D-region recombine rapidly and signal strengths return to normal.

Protons: Polar Cap Absorption (PCA)

Associated with solar flares is a release of high-energy protons. These particles can hit the Earth within 15 minutes to 2 hours of the solar flare. The protons spiral around and down the magnetic field lines of the Earth and penetrate into the atmosphere near the magnetic poles increasing the ionization of the D and E layers. PCA's typically last anywhere from about an hour to several days, with an average of around 24 to 36 hours.

Geomagnetic Storms

A geomagnetic storm is a temporary intense disturbance of the Earth's magnetosphere.

- During a geomagnetic storm the F_2 layer will become unstable, fragment, and may even disappear completely.

- In the Northern and Southern pole regions of the Earth aurorae will be observable in the sky.

Lightning

Lightning can cause ionospheric perturbations in the D-region in one of two ways. The first is through VLF (very low frequency) radio waves launched into the magnetosphere. These so-called "whistler" mode waves can interact with radiation belt particles and cause them to precipitate onto the ionosphere, adding ionization to the D-region. These disturbances are called "lightning-induced electron precipitation" (LEP) events.

Additional ionization can also occur from direct heating/ionization as a result of huge motions of charge in lightning strikes. These events are called early/fast.

In 1925, C. T. R. Wilson proposed a mechanism by which electrical discharge from lightning storms could propagate upwards from clouds to the ionosphere. Around the same time, Robert Watson-Watt, working at the Radio Research Station in Slough, UK, suggested that the ionospheric sporadic E layer (E_s) appeared to be enhanced as a result of lightning but that more work was needed. In 2005, C. Davis and C. Johnson, working at the Rutherford Appleton Laboratory in Oxfordshire, UK, demonstrated that the E_s layer was indeed enhanced as a result of lightning activity. Their subsequent research has focused on the mechanism by which this process can occur.

Applications

Radio Communication

DX communication, popular among amateur radio enthusiasts, is a term given to communication over great distances. Thanks to the property of ionized atmospheric gases to refract high frequency (HF, or shortwave) radio waves, the ionosphere can be utilized to "bounce" a transmitted signal down to ground. Transcontinental HF-connections rely on up to 5 bounces, or hops. Such communications played an important role during World War II. Karl Rawer's most sophisticated prediction methodtook account of several (zig-zag) paths, attenuation in the D-region and predicted the 11-year solar cycle by a method due to Wolfgang Gleißberg.

Mechanism of Refraction

When a radio wave reaches the ionosphere, the electric field in the wave forces the electrons in the ionosphere into oscillation at the same frequency as the radio wave. Some of the radio-frequency energy is given up to this resonant oscillation. The oscillating electrons will then either be lost to recombination or will re-radiate the original wave energy. Total refraction can occur when the collision frequency of the ionosphere is less than the radio frequency, and if the electron density in the ionosphere is great enough.

The critical frequency is the limiting frequency at or below which a radio wave is reflected by an ionospheric layer at vertical incidence. If the transmitted frequency is higher than the plasma frequency of the ionosphere, then the electrons cannot respond fast enough, and they are not able to re-radiate the signal. It is calculated as shown below:

$$f_{critical} = 9 \times \sqrt{N}$$

where N = electron density per m and $f_{critical}$ is in Hz.

The Maximum Usable Frequency (MUF) is defined as the upper frequency limit that can be used for transmission between two points at a specified time.

$$f_{\text{muf}} = \frac{f_{\text{critical}}}{\sin \alpha}$$

where α = angle of attack, the angle of the wave relative to the horizon, and sin is the sine function.

The cutoff frequency is the frequency below which a radio wave fails to penetrate a layer of the ionosphere at the incidence angle required for transmission between two specified points by refraction from the layer.

Other Applications

The open system electrodynamic tether, which uses the ionosphere, is being researched. The space tether uses plasma contactors and the ionosphere as parts of a circuit to extract energy from the Earth's magnetic field by electromagnetic induction.

Measurements

Overview

Scientists also are exploring the structure of the ionosphere by a wide variety of methods, including passive observations of optical and radio emissions generated in the ionosphere, bouncing radio waves of different frequencies from it, incoherent scatter radars such as the EISCAT, Sondre Stromfjord, Millstone Hill, Arecibo, and Jicamarca radars, coherent scatter radars such as the Super Dual Auroral Radar Network (SuperDARN) radars, and using special receivers to detect how the reflected waves have changed from the transmitted waves.

A variety of experiments, such as HAARP (High Frequency Active Auroral Research Program), involve high power radio transmitters to modify the properties of the ionosphere. These investigations focus on studying the properties and behavior of ionospheric plasma, with particular emphasis on being able to understand and use it to enhance communications and surveillance systems for both civilian and military purposes. HAARP was started in 1993 as a proposed twenty-year experiment, and is currently active near Gakona, Alaska.

The SuperDARN radar project researches the high- and mid-latitudes using coherent backscatter of radio waves in the 8 to 20 MHz range. Coherent backscatter is similar to Bragg scattering in crystals and involves the constructive interference of scattering from ionospheric density irregularities. The project involves more than 11 different countries and multiple radars in both hemispheres.

Scientists are also examining the ionosphere by the changes to radio waves, from satellites and stars, passing through it. The Arecibo radio telescope located in Puerto Rico, was originally intended to study Earth's ionosphere.

Ionograms

Ionograms show the virtual heights and critical frequencies of the ionospheric layers and which are measured by an ionosonde. An ionosonde sweeps a range of frequencies, usually from 0.1 to

30 MHz, transmitting at vertical incidence to the ionosphere. As the frequency increases, each wave is refracted less by the ionization in the layer, and so each penetrates further before it is reflected. Eventually, a frequency is reached that enables the wave to penetrate the layer without being reflected. For ordinary mode waves, this occurs when the transmitted frequency just exceeds the peak plasma, or critical, frequency of the layer. Tracings of the reflected high frequency radio pulses are known as ionograms. Reduction rules are given in: "URSI Handbook of Ionogram Interpretation and Reduction", edited by William Roy Piggott and Karl Rawer, Elsevier Amsterdam, 1961.

Incoherent Scatter Radars

Incoherent scatter radars operate above the critical frequencies. Therefore, the technique allows to probe the ionosphere, unlike ionosondes, also above the electron density peaks. The thermal fluctuations of the electron density scattering the transmitted signals lack coherence, which gave the technique its name. Their power spectrum contains information not only on the density, but also on the ion and electron temperatures, ion masses and drift velocities.

Solar Flux

Solar flux is a measurement of the intensity of solar radio emissions at a frequency of 2800 MHz made using a radio telescope located in Dominion Radio Astrophysical Observatory, Penticton, British Columbia, Canada.Known also as the 10.7 cm flux (the wavelength of the radio signals at 2800 MHz), this solar radio emission has been shown to be proportional to sunspot activity. However, the level of the Sun's ultraviolet and X-ray emissions is primarily responsible for causing ionization in the Earth's upper atmosphere. We now have data from the GOES spacecraft that measures the background X-ray flux from the Sun, a parameter more closely related to the ionization levels in the ionosphere.

- The *A* and *K* indices are a measurement of the behavior of the horizontal component of the geomagnetic field. The *K* index uses a scale from 0 to 9 to measure the change in the horizontal component of the geomagnetic field. A new *K* index is determined at the Boulder Geomagnetic Observatory.

- The geomagnetic activity levels of the Earth are measured by the fluctuation of the Earth's magnetic field in SI units called teslas (or in non-SI gauss, especially in older literature). The Earth's magnetic field is measured around the planet by many observatories. The data retrieved is processed and turned into measurement indices. Daily measurements for the entire planet are made available through an estimate of the *ap* index, called the *planetary A-index* (PAI).

GPS/GNSS

Ionospheres of Other Planets and Natural Satellites

Objects in the Solar System that have appreciable atmospheres (i.e., all of the major planets and many of the larger natural satellites) generally produce ionospheres. Planets known to have ionospheres include Venus, Uranus, Mars and Jupiter.

The atmosphere of Titan includes an ionosphere that ranges from about 1,100 km (680 mi) to 1,300 km (810 mi) in altitude and contains carbon compounds.

History

As early as 1839, the German mathematician and physicist Carl Friedrich Gauss postulated that an electrically conducting region of the atmosphere could account for observed variations of Earth's magnetic field. Sixty years later, Guglielmo Marconi received the first trans-Atlantic radio signal on December 12, 1901, in St. John's, Newfoundland (now in Canada) using a 152.4 m (500 ft) kite-supported antenna for reception. The transmitting station in Poldhu, Cornwall, used a spark-gap transmitter to produce a signal with a frequency of approximately 500 kHz and a power of 100 times more than any radio signal previously produced. The message received was three dits, the Morse code for the letter S. To reach Newfoundland the signal would have to bounce off the ionosphere twice. Dr. Jack Belrose has contested this, however, based on theoretical and experimental work.However, Marconi did achieve transatlantic wireless communications in Glace Bay, Nova Scotia, one year later.

In 1902, Oliver Heaviside proposed the existence of the *Kennelly–Heaviside layer* of the ionosphere which bears his name. Heaviside's proposal included means by which radio signals are transmitted around the Earth's curvature. Heaviside's proposal, coupled with Planck's law of black body radiation, may have hampered the growth of radio astronomy for the detection of electromagnetic waves from celestial bodies until 1932 (and the development of high-frequency radio transceivers). Also in 1902, Arthur Edwin Kennelly discovered some of the ionosphere's radio-electrical properties.

In 1912, the U.S. Congress imposed the Radio Act of 1912 on amateur radio operators, limiting their operations to frequencies above 1.5 MHz (wavelength 200 meters or smaller). The government thought those frequencies were useless. This led to the discovery of HF radio propagation via the ionosphere in 1923.

In 1926, Scottish physicist Robert Watson-Watt introduced the term *ionosphere* in a letter published only in 1969 in *Nature*:

We have in quite recent years seen the universal adoption of the term 'stratosphere'..and..the companion term 'troposphere'... The term 'ionosphere', for the region in which the main characteristic is large scale ionisation with considerable mean free paths, appears appropriate as an addition to this series.

Edward V. Appleton was awarded a Nobel Prize in 1947 for his confirmation in 1927 of the existence of the ionosphere. Lloyd Berkner first measured the height and density of the ionosphere. This permitted the first complete theory of short-wave radio propagation. Maurice V. Wilkes and J. A. Ratcliffe researched the topic of radio propagation of very long radio waves in the ionosphere. Vitaly Ginzburg has developed a theory of electromagnetic wave propagation in plasmas such as the ionosphere.

In 1962, the Canadian satellite Alouette 1 was launched to study the ionosphere. Following its success were Alouette 2 in 1965 and the two ISIS satellites in 1969 and 1971, further AEROS-A and -B in 1972 and 1975, all for measuring the ionosphere.

Stratosphere

The stratosphere is the second major layer of Earth's atmosphere, just above the troposphere, and below the mesosphere. About 20% of the atmosphere's mass is contained in the stratosphere. The stratosphere is stratified in temperature, with warmer layers higher and cooler layers closer to the Earth. The increase of temperature with altitude, is a result of the absorption of the Sun's ultraviolet radiation by ozone. This is in contrast to the troposphere, near the Earth's surface, where temperatures decreases with altitude. The border between the troposphere and stratosphere, the tropopause, marks where this temperature inversion begins. Near the equator, the stratosphere starts at 18 km (59,000 ft; 11 mi); at mid latitudes, it starts at 10–13 km (33,000–43,000 ft; 6.2–8.1 mi) and ends at 50 km (160,000 ft; 31 mi); at the poles, it starts at about 8 km (26,000 ft; 5.0 mi). Temperatures vary within the stratosphere with the seasons, in particular with the polar night (winter). The greatest variation of temperature, takes place over the poles in the lower stratosphere; those variations are largely steady at lower latitudes and higher altitudes.

Ozone and Temperature

Within this layer, temperature increases with altitude; the top of the stratosphere has a temperature of about 270 K (−3°C or 26.6°F).The increase of temperature within the stratosphere with altitude is due to the absorption of high energy ultraviolet (UVB and UVC) radiation from the Sun and the breaking of ozone (O_3) into the atomic oxygen (O_1) and common molecular oxygen (O_2), hence, greater heating. At mid-stratosphere, there is less UV light, and as a result, less radiation energy able to break up ozone. As a result, O and O_2 combine to O_3, producing ozone at the lowest level of the stratosphere. It is the making and breaking of ozone that creates the stratification of temperatures in the stratosphere and protects life on Earth from the harmful effects of ultra-violet radiation. The result is an increase of temperature with altitude, despite the constant radiation of energy to space. The lower stratosphere receives very little UVC; thus atomic oxygen is not found here and ozone is not formed. This vertical stratification, with warmer layers above and cooler layers below, makes the stratosphere dynamically stable: there is no regular convection and associated turbulence in this part of the atmosphere. However, exceptionally energetic convection processes, such as volcanic eruption columns and overshooting tops in severe supercell thunderstorms, may carry convection into the stratosphere on a very local and temporary basis.

The top of the stratosphere is called the stratopause, above which the temperature decreases with height.

Methane (CH_4), while not a direct cause of ozone destruction in the stratosphere, does lead to the formation of compounds that destroy ozone. Monatomic oxygen (O) in the upper stratosphere reacts with methane (CH_4) to form a hydroxyl radical (OH·). This hydroxyl radical is then able to interact with non-soluble compounds like chlorofluorocarbons, and UV light breaks off chlorine radicals (Cl·). These chlorine radicals break off an oxygen atom from the ozone molecule, creating an oxygen molecule (O_2) and a hypochloryl radical (ClO·). The hypochloryl radical then reacts with an atomic oxygen creating another oxygen molecule and another chlorine radical, thereby preventing the reaction of monatomic oxygen with O_2 to create natural ozone.

Aircraft Flight

Commercial airliners typically cruise at altitudes of 9–12 km (30,000–39,000 ft) which is in the lower reaches of the stratosphere in temperate latitudes.This optimizes fuel efficiency, mostly due to the low temperatures encountered near the tropopause and low air density, reducing parasitic drag on the airframe. Stated another way, it allows the airliner to fly faster for the same amount of drag. It also allows them to stay above the turbulent weather of the troposphere.

The Concorde aircraft cruised at mach 2 at about 18,000 m (59,000 ft), and the SR-71 cruised at mach 3 at 26,000 m (85,000 ft), all within the stratosphere.

Because the temperature in the tropopause and lower stratosphere is largely constant with increasing altitude, very little convection and its resultant turbulence occurs there. Most turbulence at this altitude is caused by variations in the jet stream and other local wind shears, although areas of significant convective activity (thunderstorms) in the troposphere below may produce turbulence as a result of convective overshoot.

On October 24, 2014, Alan Eustace became the record holder for reaching the altitude record for a manned balloon at 135,890 ft (41,419 m). Mr Eustace also broke the world records for vertical speed reached with a peak velocity of 1,321 km/h (822 mph) and total freefall distance of 123,414 ft (37,617 m) - lasting four minutes and 27 seconds.

Circulation and Mixing

The stratosphere is a region of intense interactions among radiative, dynamical, and chemical processes, in which the horizontal mixing of gaseous components proceeds much more rapidly than does vertical mixing.

An interesting feature of stratospheric circulation is the quasi-biennial oscillation (QBO) in the tropical latitudes, which is driven by gravity waves that are convectively generated in the troposphere. The QBO induces a secondary circulation that is important for the global stratospheric transport of tracers, such as ozone or water vapor.

During northern hemispheric winters, sudden stratospheric warmings, caused by the absorption of Rossby waves in the stratosphere, can be observed in approximately half of winters when easterly winds develop in the stratosphere. These events often precede unusual winter weather and may even be responsible for the cold European winters of the 1960s.

Stratospheric warming of the polar vortex results in its weakening. When the vortex is strong, it keeps the cold, high pressure air masses "contained" in the arctic; when the vortex weakens, air masses move equatorward, and results in rapid changes of weather in the mid latitudes.

Life

Bacteria

Bacterial life survives in the stratosphere, making it a part of the biosphere.In 2001 an Indian experiment, involving a high-altitude balloon, was carried out at a height of 41 kilometres and a sample of dust was collected with bacterial material inside.

Birds

Some bird species have been reported to fly at the lower levels of the stratosphere. On November 29, 1973, a Rüppell's vulture was ingested into a jet engine 11,552 m (37,900 ft) above the Ivory Coast, and bar-headed geese reportedly overfly Mount Everest's summit, which is 8,848 m (29,029 ft).

Discovery

Léon Teisserenc de Bort from France and Richard Assmann from Germany, in separate publications and following years of observations, announced the discovery of an isothermal layer at around 11–14 km, which is the base of the lower stratosphere. This was based on temperature profiles from unmanned instrumented balloons.

Troposphere

The troposphere is the lowest portion of Earth's atmosphere, and is also where all weather takes place. It contains approximately 75% of the atmosphere's mass and 99% of its water vapor and aerosols.The average depths of the troposphere are 20 km (12 mi) in the tropics, 17 km (11 mi) in the mid latitudes, and 7 km (4.3 mi) in the polar regions in winter. The lowest part of the troposphere, where friction with the Earth's surface influences air flow, is the planetary boundary layer. This layer is typically a few hundred meters to 2 km (1.2 mi) deep depending on the landform and time of day. Atop the troposphere is the tropopause, which is the border between the troposphere and stratosphere. The tropopause is an inversion layer, where the air temperature ceases to decline with height and remains constant through its thickness.

The word troposphere derives from the Greek: *tropos* for "turn, turn toward, trope" and "-sphere" (as in, the Earth), reflecting the fact that rotational turbulent mixing plays an important role in the troposphere's structure and behaviour. Most of the phenomena we associate with day-to-day weather occur in the troposphere.

Pressure and Temperature Structure

A view of Earth's troposphere from an airplane.

Composition

The chemical composition of the troposphere is essentially uniform, with the notable exception of water vapor. The source of water vapor is at the surface through the processes of evaporation. The temperature of the troposphere decreases with height, and saturation vapor pressure decreases strongly as temperature drops, so the amount of water vapor that can exist in the atmosphere decreases strongly with height. Thus the proportion of water vapor is normally greatest near the surface and decreases with height.

Pressure

The pressure of the atmosphere is maximum at sea level and decreases with altitude. This is because the atmosphere is very nearly in hydrostatic equilibrium, so that the pressure is equal to the weight of air above a given point. The change in pressure with height, therefore can be equated to the density with this hydrostatic equation:

$$\frac{dP}{dz} = -\rho g_n = -\frac{mPg_n}{RT}$$

where:

- g_n is the standard gravity
- ρ is the density
- z is the altitude
- P is the pressure
- R is the gas constant
- T is the thermodynamic (absolute) temperature
- m is the molar mass

Since temperature in principle also depends on altitude, one needs a second equation to determine the pressure as a function of height, as discussed in the next section.

Temperature

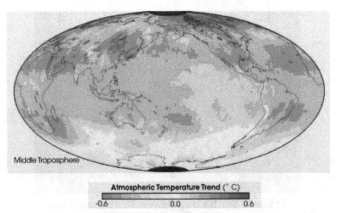

Middle Troposphere

Atmospheric Temperature Trend (° C)

-0.6   0.0   0.6

This image shows the temperature trend in the Middle Troposphere as measured by a series of satellite-based instruments between January 1979 and December 2005. The middle troposphere is centered around 5 kilometers above the surface. Oranges and yellows dominate the troposphere image, indicating that the air nearest the Earth's surface warmed during the period.Source:

The temperature of the troposphere generally decreases as altitude increases. The rate at which the temperature decreases, $-dT >$, is called the environmental lapse rate (ELR). The ELR is nothing more than the difference in temperature between the surface and the tropopause divided by the height. The ELR assumes that the air is perfectly still, i.e. that there is no mixing of the layers of air from vertical convection, nor winds that would create turbulence and hence mixing of the layers of air. The reason for this temperature difference is that the ground absorbs most of the sun's energy, which then heats the lower levels of the atmosphere with which it is in contact. Meanwhile, the radiation of heat at the top of the atmosphere results in the cooling of that part of the atmosphere.

Environmental Lapse Rate (ELR)		
Altitude Region	**Lapse rate**	**Lapse Rate**
(m)	**(Kelvin/km)**	**(°F/1000 feet)**
0 - 11,000	-6.5	-3.57
11,000 - 20,000	0.0	0.0
20,000 - 32,000	1.0	0.55
32,000 - 47,000	2.8	1.54
47,000 - 51,000	0.0	0.0
51,000 - 71,000	-2.8	-1.54
71,000 - 85,000	-2.0	-1.09

The ELR assumes the atmosphere is still, but as air is heated it becomes buoyant and rises. The dry adiabatic lapse rate accounts for the effect of the expansion of dry air as it rises in the atmosphere and wet adibatic lapse rates includes the effect of the condensation of water vapor on the lapse rate.

When a parcel of air rises, it expands, because the pressure is lower at higher altitudes. As the air parcel expands, it pushes the surrounding air outward, transferring energy in the form of work from that parcel to the atmosphere. As energy transfer to a parcel of air by way of heat is very slow, it is assumed to not exchange energy by way of heat with the environment. Such a process is called an adiabatic process (no energy transfer by way of heat). Since the rising parcel of air is losing energy as it does work on the surrounding atmosphere and no energy is transferred into it as heat from the atmosphere to make up for the loss, the parcel of air is losing energy, which manifests itself as a decrease in the temperature of the air parcel. The reverse, of course, will be true for a parcel of air that is sinking and is being compressed.

Since the process of compression and expansion of an air parcel can be considered reversible and no energy is transferred into or out of the parcel, such a process is considered isentropic, meaning that there is no change in entropy as the air parcel rises and falls, $dS = 0$. Since the heat exchanged $dQ = 0$ is related to the entropy change dS by $dQ = TdS$, the equation governing the temperature as a function of height for a thoroughly mixed atmosphere is

$$\frac{dS}{dz} = 0$$

where S is the entropy. The above equation states that the entropy of the atmosphere does not change with height. The rate at which temperature decreases with height under such conditions is called the adiabatic lapse rate.

For *dry* air, which is approximately an ideal gas, we can proceed further. The adiabatic equation for an ideal gas is

$$p(z)T(z)^{-\frac{\gamma}{\gamma-1}} = constant$$

where γ is the heat capacity ratio $\gamma =7/5$, for air). Combining with the equation for the pressure, one arrives at the dry adiabatic lapse rate,

$$\frac{dT}{dz} = -\frac{mg}{R}\frac{\gamma-1}{\gamma} = -9.8°C/km$$

If the air contains water vapor, then cooling of the air can cause the water to condense, and the behavior is no longer that of an ideal gas. If the air is at the saturated vapor pressure, then the rate at which temperature drops with height is called the saturated adiabatic lapse rate. More generally, the actual rate at which the temperature drops with altitude is called the environmental lapse rate. In the troposphere, the average environmental lapse rate is a drop of about 6.5 °C for every 1 km (1,000 meters) in increased height.

The environmental lapse rate (the actual rate at which temperature drops with height, dT/dz) is not usually equal to the adiabatic lapse rate (or correspondingly, $dS/dz \neq 0$). If the upper air is warmer than predicted by the adiabatic lapse rate $dT/dz > 0$), then when a parcel of air rises and expands, it will arrive at the new height at a lower temperature than its surroundings. In this case, the air parcel is denser than its surroundings, so it sinks back to its original height, and the air is stable against being lifted. If, on the contrary, the upper air is cooler than predicted by the adiabatic lapse rate, then when the air parcel rises to its new height it will have a higher temperature and a lower density than its surroundings, and will continue to accelerate upward.

The troposphere is heated from below by latent heat, longwave radiation, and sensible heat. Surplus heating and vertical expansion of the troposphere occurs in the tropics. At middle latitudes, tropospheric temperatures decrease from an average of 15 °C at sea level to about -55 °C at the tropopause. At the poles, tropospheric temperature only decreases from an average of 0 °C at sea level to about -45 °C at the tropopause. At the equator, tropospheric temperatures decrease from an average of 20 °C at sea level to about -70 to -75 °C at the tropopause. The troposphere is thinner at the poles and thicker at the equator. The average thickness of the tropical tropopause is roughly 7 kilometers greater than the average tropopause thickness at the poles.

Tropopause

The tropopause is the boundary region between the troposphere and the stratosphere.

Measuring the temperature change with height through the troposphere and the stratosphere identifies the location of the tropopause. In the troposphere, temperature decreases with altitude. In the stratosphere, however, the temperature remains constant for a while and then increases with altitude. The region of the atmosphere where the lapse rate changes from positive (in the troposphere) to negative (in the stratosphere), is defined as the tropopause.Thus, the tropopause is an inversion layer, and there is little mixing between the two layers of the atmosphere.

Atmospheric Flow

The flow of the atmosphere generally moves in a west to east direction. This, however, can often become interrupted, creating a more north to south or south to north flow. These scenarios are often described in meteorology as zonal or meridional. These terms, however, tend to be used in reference to localised areas of atmosphere (at a synoptic scale). A fuller explanation of the flow of atmosphere around the Earth as a whole can be found in the three-cell model.

Zonal Flow

A zonal flow regime is the meteorological term meaning that the general flow pattern is west to east along the Earth's latitude lines, with weak shortwaves embedded in the flow.The use of the word "zone" refers to the flow being along the Earth's latitudinal "zones". This pattern can buckle and thus become a meridional flow.

Meridional Flow

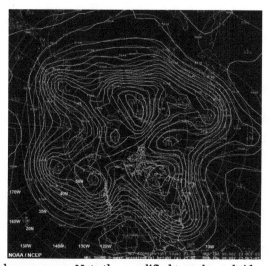

Meridional Flow pattern of October 23, 2003. Note the amplified troughs and ridges in this 500 hPa height pattern.

When the zonal flow buckles, the atmosphere can flow in a more longitudinal (or meridional) direction, and thus the term "meridional flow" arises. Meridional flow patterns feature strong, amplified troughs of low pressure and ridges of high pressure, with more north-south flow in the general pattern than west-to-east flow.

Three-cell Model

The three cells model of the atmosphere attempts to describe the actual flow of the Earth's atmosphere as a whole. It divides the Earth into the tropical (Hadley cell), mid latitude (Ferrel cell), and polar (polar cell) regions, to describe energy flow and global atmospheric circulation (mass flow). Its fundamental principle is that of balance - the energy that the Earth absorbs from the sun each year is equal to that which it loses to space by radiation. This overall Earth energy balance, however, does not apply in each latitude due to the varying strength of the sun in each "cell" as a result of the tilt of the Earth's axis in relation to its orbit. The result is a circulation of the atmosphere that transports warm air poleward from the tropics and cold air equatorward from the poles. The effect of the three cells is the tendency to even out the heat and moisture in the Earth's atmosphere around the planet.

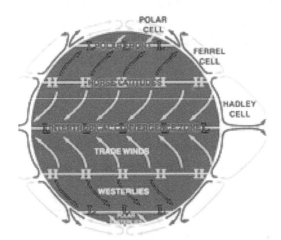

Atmospheric circulation shown with three large cells.

Synoptic Scale Observations and Concepts

Forcing Term by Meteorologists

Forcing is a term used by meteorologists to describe the situation where a change or an event in one part of the atmosphere causes a strengthening change in another part of the atmosphere. It is usually used to describe connections between upper, middle or lower levels (such as upper-level divergence causing lower level convergence in cyclone formation), but also be to describe such connections over lateral distance rather than height alone. In some respects, teleconnections could be considered a type of forcing.

Divergence and Convergence

An area of convergence is one in which the total mass of air is increasing with time, resulting in an increase in pressure at locations below the convergence level (recall that atmospheric pressure is just the total weight of air above a given point). Divergence is the opposite of convergence - an area where the total mass of air is decreasing with time, resulting in falling pressure in regions below the area of divergence. Where divergence is occurring in the upper atmosphere, there will be air coming in to try to balance the net loss of mass (this is called the principle of mass conservation), and there is a resulting upward motion (positive vertical velocity). Another way to state this is to say that regions of upper air divergence are conducive to lower level convergence, cyclone formation, and positive vertical velocity. Therefore, identifying regions of upper air divergence is an important step in forecasting the formation of a surface low pressure area.

Atmospheric Chemistry

Atmospheric chemistry is a branch of atmospheric science in which the chemistry of the Earth's atmosphere and that of other planets is studied. It is a multidisciplinary approach of research and draws on environmental chemistry, physics, meteorology, computer modeling, oceanography, geology and volcanology and other disciplines. Research is increasingly connected with other arenas of study such as climatology.

The composition and chemistry of the Earth's atmosphere is of importance for several reasons, but primarily because of the interactions between the atmosphere and living organisms. The composition of the Earth's atmosphere changes as result of natural processes such as volcano emissions, lightning and bombardment by solar particles from corona. It has also been changed by human activity and some of these changes are harmful to human health, crops and ecosystems. Examples of problems which have been addressed by atmospheric chemistry include acid rain, ozone depletion, photochemical smog, greenhouse gases and global warming. Atmospheric chemists seek to understand the causes of these problems, and by obtaining a theoretical understanding of them, allow possible solutions to be tested and the effects of changes in government policy evaluated.

Atmospheric Composition

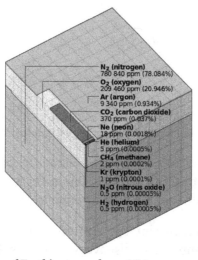

Visualisation of composition by volume of Earth's atmosphere. Water vapour is not included as it is highly variable. Each tiny cube (such as the one representing krypton) has one millionth of the volume of the entire block. Data is from NASA Langley.

Average composition of dry atmosphere (mole fractions)		
Gas	per NASA	
Nitrogen, N_2	78.084%	
Oxygen, O_2	20.946%	
Minor constituents (mole fractions in ppm)		
Argon, Ar	9340	
Carbon dioxide, CO_2	400	
Neon, Ne	18.18	
Helium, He	5.24	
Methane, CH_4	1.7	
Krypton, Kr	1.14	

Hydrogen, H_2	0.55	
Nitrous oxide, N_2O	0.5	
Xenon, Xe	0.09	
Nitrogen dioxide, NO_2	0.02	
Water		
Water vapour	Highly variable; typically makes up about 1%	

Notes: the concentration of CO_2 and CH_4 vary by season and location. The mean molecular mass of air is 28.97 g/mol. Ozone (O_3) is not included due to its high variability.

History

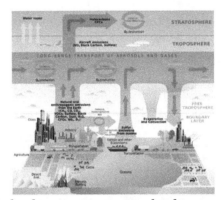

Schematic of chemical and transport processes related to atmospheric composition.

The ancient Greeks regarded air as one of the four elements, but the first scientific studies of atmospheric composition began in the 18th century. Chemists such as Joseph Priestley, Antoine Lavoisier and Henry Cavendish made the first measurements of the composition of the atmosphere.

In the late 19th and early 20th centuries interest shifted towards trace constituents with very small concentrations. One particularly important discovery for atmospheric chemistry was the discovery of ozone by Christian Friedrich Schönbein in 1840.

In the 20th century atmospheric science moved on from studying the composition of air to a consideration of how the concentrations of trace gases in the atmosphere have changed over time and the chemical processes which create and destroy compounds in the air. Two particularly important examples of this were the explanation by Sydney Chapman and Gordon Dobson of how the ozone layer is created and maintained, and the explanation of photochemical smog by Arie Jan Haagen-Smit. Further studies on ozone issues led to the 1995 Nobel Prize in Chemistry award shared between Paul Crutzen, Mario Molina and Frank Sherwood Rowland.

In the 21st century the focus is now shifting again. Atmospheric chemistry is increasingly studied as one part of the Earth system. Instead of concentrating on atmospheric chemistry in isolation the focus is now on seeing it as one part of a single system with the rest of the atmosphere, biosphere and geosphere. An especially important driver for this is the links between chemistry and climate such as the effects of changing climate on the recovery of the ozone hole and vice versa but also interaction of the composition of the atmosphere with the oceans and terrestrial ecosystems.

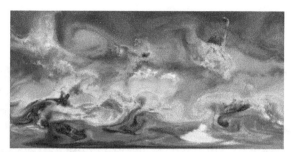

Carbon dioxide in Earth's atmosphere if *half* of global-warming emissions are *not* absorbed.
(NASA simulation; 9 November 2015)

Methodology

Observations, lab measurements and modeling are the three central elements in atmospheric chemistry. Progress in atmospheric chemistry is often driven by the interactions between these components and they form an integrated whole. For example, observations may tell us that more of a chemical compound exists than previously thought possible. This will stimulate new modelling and laboratory studies which will increase our scientific understanding to a point where the observations can be explained.

Observation

Observations of atmospheric chemistry are essential to our understanding. Routine observations of chemical composition tell us about changes in atmospheric composition over time. One important example of this is the Keeling Curve - a series of measurements from 1958 to today which show a steady rise in of the concentration of carbon dioxide. Observations of atmospheric chemistry are made in observatories such as that on Mauna Loa and on mobile platforms such as aircraft (e.g. the UK's Facility for Airborne Atmospheric Measurements), ships and balloons. Observations of atmospheric composition are increasingly made by satellites with important instruments such as GOME and MOPITT giving a global picture of air pollution and chemistry. Surface observations have the advantage that they provide long term records at high time resolution but are limited in the vertical and horizontal space they provide observations from. Some surface based instruments e.g. LIDAR can provide concentration profiles of chemical compounds and aerosol but are still restricted in the horizontal region they can cover. Many observations are available on line in Atmospheric Chemistry Observational Databases.

Lab Measurements

Measurements made in the laboratory are essential to our understanding of the sources and sinks of pollutants and naturally occurring compounds. Lab studies tell us which gases react with each other and how fast they react. Measurements of interest include reactions in the gas phase, on surfaces and in water. Also of high importance is photochemistry which quantifies how quickly molecules are split apart by sunlight and what the products are plus thermodynamic data such as Henry's law coefficients.

Modeling

In order to synthesise and test theoretical understanding of atmospheric chemistry, computer models (such as chemical transport models) are used. Numerical models solve the differential equations governing the concentrations of chemicals in the atmosphere. They can be very simple or very complicated. One common trade off in numerical models is between the number of chemical compounds and chemical reactions modelled versus the representation of transport and mixing in the atmosphere. For example, a box model might include hundreds or even thousands of chemical reactions but will only have a very crude representation of mixing in the atmosphere. In contrast, 3D models represent many of the physical processes of the atmosphere but due to constraints on computer resources will have far fewer chemical reactions and compounds. Models can be used to interpret observations, test understanding of chemical reactions and predict future concentrations of chemical compounds in the atmosphere. One important current trend is for atmospheric chemistry modules to become one part of earth system models in which the links between climate, atmospheric composition and the biosphere can be studied.

Some models are constructed by automatic code generators (e.g. Autochem or KPP). In this approach a set of constituents are chosen and the automatic code generator will then select the reactions involving those constituents from a set of reaction databases. Once the reactions have been chosen the ordinary differential equations (ODE) that describe their time evolution can be automatically constructed.

Ozone Layer

The ozone layer or ozone shield is a region of Earth's stratosphere that absorbs most of the Sun's ultraviolet (UV) radiation. It contains high concentrations of ozone (O_3) in relation to other parts of the atmosphere, although still small in relation to other gases in the stratosphere. The ozone layer contains less than 10 parts per million of ozone, while the average ozone concentration in Earth's atmosphere as a whole is about 0.3 parts per million. The ozone layer is mainly found in the lower portion of the stratosphere, from approximately 20 to 30 kilometres (12 to 19 mi) above Earth, although the thickness varies seasonally and geographically.

The ozone layer was discovered in 1913 by the French physicists Charles Fabry and Henri Buisson. Measurements of the sun showed that the radiation sent out from its surface and reaching the ground on Earth is usually consistent with the spectrum of a black body with a temperature in the range of 5,500–6,000 K (5,227 to 5,727 °C), except that there was no radiation below a wavelength of about 310 nm at the ultraviolet end of the spectrum. It was deduced that the missing radiation was being absorbed by something in the atmosphere. Eventually the spectrum of the missing radiation was matched to only one known chemical, ozone.Its properties were explored in detail by the British meteorologist G. M. B. Dobson, who developed a simple spectrophotometer (the Dobsonmeter) that could be used to measure stratospheric ozone from the ground. Between 1928 and 1958, Dobson established a worldwide network of ozone monitoring stations, which continue to operate to this day. The "Dobson unit", a convenient measure of the amount of ozone overhead, is named in his honor.

The ozone layer absorbs 97 to 99 percent of the Sun's medium-frequency ultraviolet light (from about 200 nm to 315 nm wavelength), which otherwise would potentially damage exposed life forms near the surface.

The United Nations General Assembly has designated September 16 as the International Day for the Preservation of the Ozone Layer.

Venus also has a thin ozone layer at an altitude of 100 kilometers from the planet's surface.

Sources

The photochemical mechanisms that give rise to the ozone layer were discovered by the British physicist Sydney Chapman in 1930. Ozone in the Earth's stratosphere is created by ultraviolet light striking ordinary oxygen molecules containing two oxygen atoms (O_2), splitting them into individual oxygen atoms (atomic oxygen); the atomic oxygen then combines with unbroken O_2 to create ozone, O_3. The ozone molecule is unstable (although, in the stratosphere, long-lived) and when ultraviolet light hits ozone it splits into a molecule of O_2 and an individual atom of oxygen, a continuing process called the ozone-oxygen cycle. Chemically, this can be described as:

$$O_2 + h\nu_{uv} \; 2O$$

$$O + O_2 \leftrightarrow O_3$$

About 90 percent of the ozone in our atmosphere is contained in the stratosphere. Ozone concentrations are greatest between about 20 and 40 kilometres (66,000 and 131,000 ft), where they range from about 2 to 8 parts per million. If all of the ozone were compressed to the pressure of the air at sea level, it would be only 3 millimetres ($\frac{1}{8}$ inch) thick.

Ultraviolet Light

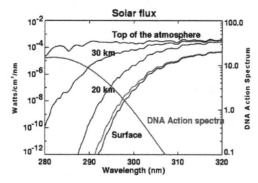

UV-B energy levels at several altitudes. Blue line shows DNA sensitivity. Red line shows surface energy level with 10 percent decrease in ozone

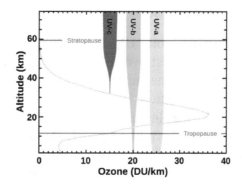

Levels of ozone at various altitudes and blocking of different bands of ultraviolet radiation. Essentially all UVC (100–280 nm) is blocked by dioxygen (from 100–200 nm) or else by ozone (200–280 nm) in the atmosphere. The shorter portion of the UV-C band and the more energetic UV above this band causes the formation of the ozone layer, when single oxygen atoms produced by UV photolysis of dioxygen (below 240 nm) react with more dioxygen. The ozone layer also blocks most, but not quite all, of the sunburn-producing UV-B (280–315 nm) band, which lies in the wavelengths longer than UV-C. The band of UV closest to visible light, UV-A (315–400 nm), is hardly affected by ozone, and most of it reaches the ground. UV-A does not primarily cause skin reddening, but there is evidence that it causes long-term skin damage.

Although the concentration of the ozone in the ozone layer is very small, it is vitally important to life because it absorbs biologically harmful ultraviolet (UV) radiation coming from the sun. Extremely short or vacuum UV (10–100 nm) is screened out by nitrogen. UV radiation capable of penetrating nitrogen is divided into three categories, based on its wavelength; these are referred to as UV-A (400–315 nm), UV-B (315–280 nm), and UV-C (280–100 nm).

UV-C, which is very harmful to all living things, is entirely screened out by a combination of dioxygen (< 200 nm) and ozone (> about 200 nm) by around 35 kilometres (115,000 ft) altitude. UV-B radiation can be harmful to the skin and is the main cause of sunburn; excessive exposure can also cause cataracts, immune system suppression, and genetic damage, resulting in problems such as skin cancer. The ozone layer (which absorbs from about 200 nm to 310 nm with a maximal absorption at about 250 nm)is very effective at screening out UV-B; for radiation with a wavelength of 290 nm, the intensity at the top of the atmosphere is 350 million times stronger than at the Earth's surface. Nevertheless, some UV-B, particularly at its longest wavelengths, reaches the surface, and is important for the skin's production of vitamin D.

Ozone is transparent to most UV-A, so most of this longer-wavelength UV radiation reaches the surface, and it constitutes most of the UV reaching the Earth. This type of UV radiation is significantly less harmful to DNA, although it may still potentially cause physical damage, premature aging of the skin, indirect genetic damage, and skin cancer.

Distribution in the Stratosphere

The thickness of the ozone layer—that is, the total amount of ozone in a column overhead—varies by a large factor worldwide, being in general smaller near the equator and larger towards the poles. It also varies with season, being in general thicker during the spring and thinner during the autumn. The reasons for this latitude and seasonal dependence are complicated, which involve in atmospheric circulation patterns as well as solar intensity.

Since stratospheric ozone is produced by solar UV radiation, one might expect to find the highest ozone levels over the tropics and the lowest over polar regions. The same argument would lead one to expect the highest ozone levels in the summer and the lowest in the winter. The observed behavior is very different: most of the ozone is found in the mid-to-high latitudes of the northern and southern hemispheres, and the highest levels are found in the spring, not summer, and the lowest in the autumn, not winter in the northern hemisphere. During winter, the ozone layer actually increases in depth. This puzzle is explained by the prevailing stratospheric wind patterns, known as the Brewer-Dobson circulation. While most of the ozone is indeed created over the tropics, the stratospheric circulation then transports it poleward and downward to the lower stratosphere of the high latitudes.However, owing to the ozone hole phenomenon, the lowest amounts of column ozone found anywhere in the world are over the Antarctic in the southern spring period of September and October and to a lesser extent over the Arctic in the northern spring period of March, April, and May.

The ozone layer is higher in altitude in the tropics, and lower in altitude outside the tropics, especially in the polar regions. This altitude variation of ozone results from the slow circulation that lifts the ozone-poor air out of the troposphere into the stratosphere. As this air slowly rises in the tropics, ozone is produced as the sun overhead photolyzes oxygen molecules. As this slow circulation levels off and flows towards the mid-latitudes, it carries the ozone-rich air from the tropical middle stratosphere to the lower stratosphere middle and high latitudes . The high ozone concentrations at high latitudes are due to the accumulation of ozone at lower altitudes.

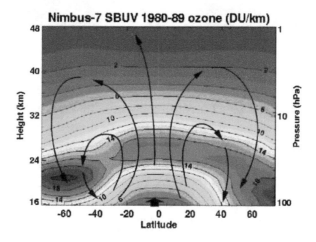

Brewer-Dobson circulation in the ozone layer.

The Brewer-Dobson circulation moves very slowly. The time needed to lift an air parcel by 1 km in the lower tropical stratosphere is about 2 months (18 m per day). However, horizontal poleward transport in the lower stratosphere is much faster and amounts to approximately 100 km per day in the northern hemisphere whilst it is only half as much in the southern hemisphere (~51 km per day).Even though ozone in the lower tropical stratosphere is produced at a very slow rate, the lifting circulation is so slow that ozone can build up to relatively high levels by the time it reaches 26 kilometres (16 mi).

Ozone amounts over the continental United States (25°N to 49°N) are highest in the northern spring (April and May). These ozone amounts fall over the course of the summer to their lowest amounts in October, and then rise again over the course of the winter.Again, wind transport of ozone is principally responsible for the seasonal changes of these higher latitude ozone patterns.

The total column amount of ozone generally increases as we move from the tropics to higher latitudes in both hemispheres. However, the overall column amounts are greater in the northern hemisphere high latitudes than in the southern hemisphere high latitudes. In addition, while the highest amounts of column ozone over the Arctic occur in the northern spring (March–April), the opposite is true over the Antarctic, where the lowest amounts of column ozone occur in the southern spring (September–October).

Depletion

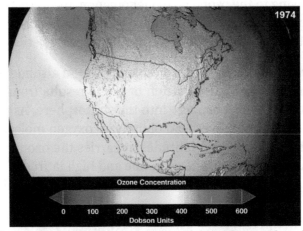

NASA projections of stratospheric ozone concentrations if chlorofluorocarbons had not been banned.

The ozone layer can be depleted by free radical catalysts, including nitric oxide (NO), nitrous oxide (N_2O), hydroxyl (OH), atomic chlorine (Cl), and atomic bromine (Br). While there are natural sources for all of these species, the concentrations of chlorine and bromine increased markedly in recent decades because of the release of large quantities of man-made organohalogen compounds, especially chlorofluorocarbons (CFCs) and bromofluorocarbons.These highly stable compounds are capable of surviving the rise to the stratosphere, where Cl and Br radicals are liberated by the action of ultraviolet light. Each radical is then free to initiate and catalyze a chain reaction capable of breaking down over 100,000 ozone molecules. By 2009, nitrous oxide was the largest ozone-depleting substance (ODS) emitted through human activities.

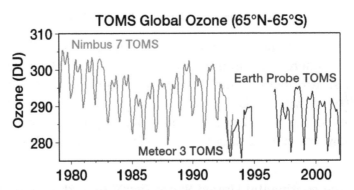

Levels of atmospheric ozone measured by satellite show clear seasonal variations and appear to verify their decline over time.

The breakdown of ozone in the stratosphere results in reduced absorption of ultraviolet radiation. Consequently, unabsorbed and dangerous ultraviolet radiation is able to reach the Earth's surface at a higher intensity. Ozone levels have dropped by a worldwide average of about 4 percent since the late 1970s. For approximately 5 percent of the Earth's surface, around the north and south poles, much larger seasonal declines have been seen, and are described as "ozone holes".The discovery of the annual depletion of ozone above the Antarctic was first announced by Joe Farman, Brian Gardiner and Jonathan Shanklin, in a paper which appeared in *Nature* on May 16, 1985.

Regulation

To support successful regulation attempts, the ozone case was communicated to lay persons "with easy-to-understand bridging metaphors derived from the popular culture" and related to "immediate risks with everyday relevance". The specific metaphors used in the discussion (ozone shield, ozone hole) proved quite useful and, compared to global climate change, the ozone case was much more seen as a "hot issue" and imminent risk. Lay people were cautious about a depletion of the ozone layer and the risks of skin cancer.

In 1978, the United States, Canada and Norway enacted bans on CFC-containing aerosol sprays that damage the ozone layer. The European Community rejected an analogous proposal to do the same. In the U.S., chlorofluorocarbons continued to be used in other applications, such as refrigeration and industrial cleaning, until after the discovery of the Antarctic ozone hole in 1985. After negotiation of an international treaty (the Montreal Protocol), CFC production was capped at 1986 levels with commitments to long-term reductions. Since that time, the treaty was amended to ban CFC production after 1995 in the developed countries, and later in developing countries. Today, all of the world's 197 countries have signed the treaty. Beginning January 1, 1996, only recycled and stockpiled CFCs were available for use in developed countries like the US. This production phaseout was possible because of efforts to ensure that there would be substitute chemicals and technologies for all ODS uses.

On August 2, 2003, scientists announced that the global depletion of the ozone layer may be slowing down because of the international regulation of ozone-depleting substances. In a study organized by the American Geophysical Union, three satellites and three ground stations confirmed that the upper-atmosphere ozone-depletion rate slowed down significantly during the previous decade. Some breakdown can be expected to continue because of ODSs used by nations which have not banned them, and because of gases which are already in the stratosphere. Some ODSs, including CFCs, have very long atmospheric lifetimes, ranging from 50 to over 100 years. It has been estimated that the ozone layer will recover to 1980 levels near the middle of the 21st century. A gradual trend toward "healing" was reported in 2016.

Compounds containing C–H bonds (such as hydrochlorofluorocarbons, or HCFCs) have been designed to replace CFCs in certain applications. These replacement compounds are more reactive and less likely to survive long enough in the atmosphere to reach the stratosphere where they could affect the ozone layer. While being less damaging than CFCs, HCFCs can have a negative impact on the ozone layer, so they are also being phased out. These in turn are being replaced by hydrofluorocarbons (HFCs) and other compounds that do not destroy stratospheric ozone at all.

Implications for Astronomy

As ozone in the atmosphere prevents most energetic ultraviolet radiation reaching the surface of the earth, astronomical data in these wavelengths has to be gathered from satellites orbiting above the atmosphere and ozone layer. Most of the light from young hot stars is in the ultraviolet and so study of these wavelengths is important for studying the origins of galaxies. The Galaxy Evolution Explorer, GALEX, is an orbiting ultraviolet space telescope launched on April 28, 2003, which operated until early 2012.

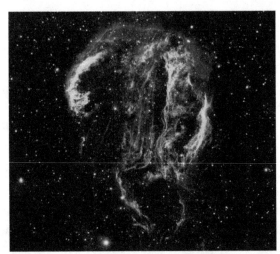

This GALEX image of the Cygnus Loop Nebula could not have been taken from the surface of the Earth because the Ozone Layer blocks the ultra-violet radiation emitted by the nebula.

References

- Wallace, John M. and Peter V. Hobbs. Atmospheric Science; An Introductory Survey.Elsevier. Second Edition, 2006. ISBN 978-0-12-732951-2.

- Jones, Daniel (2003) [1917], Peter Roach, James Hartmann and Jane Setter, eds., English Pronouncing Dictionary, Cambridge: Cambridge University Press, ISBN 3-12-539683-2.

- Tabin, Shagoon (2008). Global Warming: The Effect Of Ozone Depletion. APH Publishing. p. 194. ISBN 9788131303962. Retrieved 12 January 2016.

- Solomon, Susan, et al. (June 30, 2016). "Emergence of healing in the Antarctic ozone layer". Science. 353 (6296): 269–74. doi:10.1126/science.aae0061. PMID 27365314.

- Steigerwald, William (17 August 2015). "NASA's LADEE Spacecraft Finds Neon in Lunar Atmosphere". NASA. Retrieved 18 August 2015.

- St. Fleur, Nicholas (10 November 2015). "Atmospheric Greenhouse Gas Levels Hit Record, Report Says". New York Times. Retrieved 11 November 2015.

- Ritter, Karl (9 November 2015). "UK: In 1st, global temps average could be 1 degree C higher". AP News. Retrieved 11 November 2015.

- Cole, Steve; Gray, Ellen (14 December 2015). "New NASA Satellite Maps Show Human Fingerprint on Global Air Quality". NASA. Retrieved 14 December 2015.

- Matsumi, Y.; Kawasaki, M. (2003). "Photolysis of Atmospheric Ozone in the Ultraviolet Region" (PDF). Chem. Rev. 103 (12): 4767–4781. doi:10.1021/cr0205255. PMID 14664632. Archived from the original (PDF) on June 17, 2012. Retrieved March 14, 2015.

- Ungar, Sheldon (2000). "Knowledge, ignorance and the popular culture: climate change versus the ozone hole". Public Understanding of Science. 9 (3): 297–312. doi:10.1088/0963-6625/9/3/306. Retrieved March 14, 2015.

- Grundmann, Reiner (2007). "Climate Change and Knowledge Politics" (PDF). Environmental Politics. 16 (3): 414–432. doi:10.1080/09644010701251656. Retrieved March 14, 2015.

- Zimmer, Carl (3 October 2013). "Earth's Oxygen: A Mystery Easy to Take for Granted". New York Times. Retrieved 3 October 2013.

- Source for figures: Carbon dioxide, NOAA Earth System Research Laboratory, (updated 2013-03). Methane, IPCC TAR table 6.1, (updated to 1998). The NASA total was 17 ppmv over 100%, and CO_2 was increased here by 15 ppmv. To normalize, N_2 should be reduced by about 25 ppmv and O_2 by about 7 ppmv.

- Zimmer, Carl (3 October 2013). "Earth's Oxygen: A Mystery Easy to Take for Granted". New York Times. Retrieved 3 October 2013.

- Flury, T.; Wu, D.L.; Read, W.G. (2013). "Variability in the speed of the Brewer–Dobson circulation as observed by Aura/MLS". Atmos. Chem. Phys. 13 (9): 4563–4575. Bibcode:2013ACP....13.4563F. doi:10.5194/acp-13-4563-2013. CS1 maint: Uses authors parameter (link)Joe Buchdahl. "Atmosphere, Climate & Environment Information Programme". Ace.mmu.ac.uk. Retrieved 2012-04-18.

- Mesosphere (Wayback Machine Archive), Atmosphere, Climate & Environment Information ProgGFKDamme (UK Department for Environment, Food and Rural Affairs), archived from the original on 1 July 2010, retrieved 14 November 2011

- "Research on Balloon to Float over 50km Altitude". Institute of Space and Astronautical Science, JAXA. Retrieved 29 September 2011.

- Zahnle, K.; Schaefer, L.; Fegley, B. (2010). "Earth's Earliest Atmospheres". Cold Spring Harbor Perspectives in Biology. 2 (10): a004895. doi:10.1101/cshperspect.a004895. PMC 2944365. PMID 20573713.

- Morrisette, Peter M. (1989). "The Evolution of Policy Responses to Stratospheric Ozone Depletion". Natural Resources Journal. 29: 793–820. Retrieved 2010-04-20.

3

Key Concepts of Meteorology

This chapter provides the reader with the key concepts of meteorology like climate, weather, precipitation, air mass, atmospheric pressure, weather front, dew point and others. This helps the reader form a better understanding of these variables and their interaction. The chapter strategically encompasses and incorporates the major components and key concepts of meteorology, providing a complete understanding.

Climate

Climate is the statistics of weather, usually over a 30-year interval. It is measured by assessing the patterns of variation in temperature, humidity, atmospheric pressure, wind, precipitation, atmospheric particle count and other meteorological variables in a given region over long periods of time. Climate differs from weather, in that weather only describes the short-term conditions of these variables in a given region.

A region's climate is generated by the climate system, which has five components: atmosphere, hydrosphere, cryosphere, lithosphere, and biosphere.

The climate of a location is affected by its latitude, terrain, and altitude, as well as nearby water bodies and their currents. Climates can be classified according to the average and the typical ranges of different variables, most commonly temperature and precipitation. The most commonly used classification scheme was Köppen climate classification originally developed by Wladimir Köppen. The Thornthwaite system, in use since 1948, incorporates evapotranspiration along with temperature and precipitation information and is used in studying biological diversity and the potential effects on it of climate changes. The Bergeron and Spatial Synoptic Classification systems focus on the origin of air masses that define the climate of a region.

Paleoclimatology is the study of ancient climates. Since direct observations of climate are not available before the 19th century, paleoclimates are inferred from *proxy variables* that include non-biotic evidence such as sediments found in lake beds and ice cores, and biotic evidence such as tree rings and coral. Climate models are mathematical models of past, present and future climates. Climate change may occur over long and short timescales from a variety of factors; recent warming is discussed in global warming.

Definition

Climate is commonly defined as the weather averaged over a long period. The standard averaging period is 30 years, but other periods may be used depending on the purpose. Climate also includes statistics other than the average, such as the magnitudes of day-to-day or year-to-year variations.

The Intergovernmental Panel on Climate Change (IPCC) 2001 glossary definition is as follows:

Climate in a narrow sense is usually defined as the "average weather," or more rigorously, as the statistical description in terms of the mean and variability of relevant quantities over a period ranging from months to thousands or millions of years. The classical period is 30 years, as defined by the World Meteorological Organization (WMO). These quantities are most often surface variables such as temperature, precipitation, and wind. Climate in a wider sense is the state, including a statistical description, of the climate system.

The World Meteorological Organization (WMO) describes climate "normals" as "reference points used by climatologists to compare current climatological trends to that of the past or what is considered 'normal'. A Normal is defined as the arithmetic average of a climate element (e.g. temperature) over a 30-year period. A 30 year period is used, as it is long enough to filter out any interannual variation or anomalies, but also short enough to be able to show longer climatic trends." The WMO originated from the International Meteorological Organization which set up a technical commission for climatology in 1929. At its 1934 Wiesbaden meeting the technical commission designated the thirty-year period from 1901 to 1930 as the reference time frame for climatological standard normals. In 1982 the WMO agreed to update climate normals, and in these were subsequently completed on the basis of climate data from 1 January 1961 to 31 December 1990.

The difference between climate and weather is usefully summarized by the popular phrase "Climate is what you expect, weather is what you get." Over historical time spans there are a number of nearly constant variables that determine climate, including latitude, altitude, proportion of land to water, and proximity to oceans and mountains. These change only over periods of millions of years due to processes such as plate tectonics. Other climate determinants are more dynamic: the thermohaline circulation of the ocean leads to a 5 °C (9 °F) warming of the northern Atlantic Ocean compared to other ocean basins. Other ocean currents redistribute heat between land and water on a more regional scale. The density and type of vegetation coverage affects solar heat absorption, water retention, and rainfall on a regional level. Alterations in the quantity of atmospheric greenhouse gases determines the amount of solar energy retained by the planet, leading to global warming or global cooling. The variables which determine climate are numerous and the interactions complex, but there is general agreement that the broad outlines are understood, at least insofar as the determinants of historical climate change are concerned.

Climate Classification

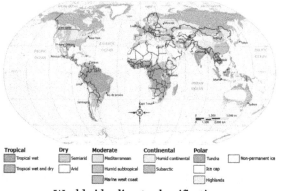

Worldwide climate classifications

There are several ways to classify climates into similar regimes. Originally, climes were defined in Ancient Greece to describe the weather depending upon a location's latitude. Modern climate classification methods can be broadly divided into *genetic* methods, which focus on the causes of climate, and *empiric* methods, which focus on the effects of climate. Examples of genetic classification include methods based on the relative frequency of different air mass types or locations within synoptic weather disturbances. Examples of empiric classifications include climate zones defined by plant hardiness, evapotranspiration, or more generally the Köppen climate classification which was originally designed to identify the climates associated with certain biomes. A common shortcoming of these classification schemes is that they produce distinct boundaries between the zones they define, rather than the gradual transition of climate properties more common in nature.

Bergeron and Spatial Synoptic

The simplest classification is that involving air masses. The Bergeron classification is the most widely accepted form of air mass classification. Air mass classification involves three letters. The first letter describes its moisture properties, with c used for continental air masses (dry) and m for maritime air masses (moist). The second letter describes the thermal characteristic of its source region: T for tropical, P for polar, A for Arctic or Antarctic, M for monsoon, E for equatorial, and S for superior air (dry air formed by significant downward motion in the atmosphere). The third letter is used to designate the stability of the atmosphere. If the air mass is colder than the ground below it, it is labeled k. If the air mass is warmer than the ground below it, it is labeled w. While air mass identification was originally used in weather forecasting during the 1950s, climatologists began to establish synoptic climatologies based on this idea in 1973.

Based upon the Bergeron classification scheme is the Spatial Synoptic Classification system (SSC). There are six categories within the SSC scheme: Dry Polar (similar to continental polar), Dry Moderate (similar to maritime superior), Dry Tropical (similar to continental tropical), Moist Polar (similar to maritime polar), Moist Moderate (a hybrid between maritime polar and maritime tropical), and Moist Tropical (similar to maritime tropical, maritime monsoon, or maritime equatorial).

Köppen

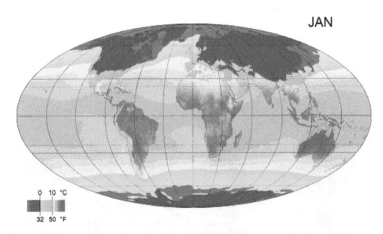

Monthly average surface temperatures from 1961–1990. This is an example of how climate varies with location and season

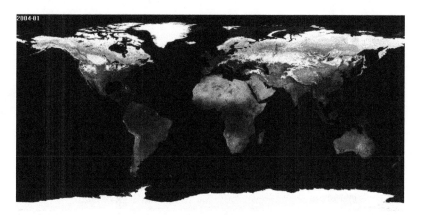

Monthly global images from NASA Earth Observatory (interactive SVG)

The Köppen classification depends on average monthly values of temperature and precipitation. The most commonly used form of the Köppen classification has five primary types labeled A through E. These primary types are A, tropical; B, dry; C, mild mid-latitude; D, cold mid-latitude; and E, polar. The five primary classifications can be further divided into secondary classifications such as rain forest, monsoon, tropical savanna, humid subtropical, humid continental, oceanic climate, Mediterranean climate, steppe, subarctic climate, tundra, polar ice cap, and desert.

Rain forests are characterized by high rainfall, with definitions setting minimum normal annual rainfall between 1,750 millimetres (69 in) and 2,000 millimetres (79 in). Mean monthly temperatures exceed 18 °C (64 °F) during all months of the year.

A monsoon is a seasonal prevailing wind which lasts for several months, ushering in a region's rainy season. Regions within North America, South America, Sub-Saharan Africa, Australia and East Asia are monsoon regimes.

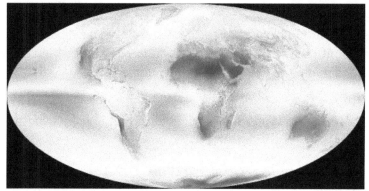

The world's cloudy and sunny spots. NASA Earth Observatory map using data collected between July 2002 and April 2015.

A tropical savanna is a grassland biome located in semiarid to semi-humid climate regions of subtropical and tropical latitudes, with average temperatures remain at or above 18 °C (64 °F) year round and rainfall between 750 millimetres (30 in) and 1,270 millimetres (50 in) a year. They are widespread on Africa, and are found in India, the northern parts of South America, Malaysia, and Australia.

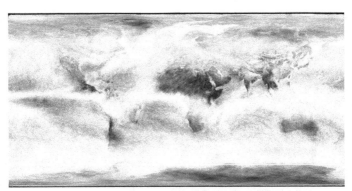

Cloud cover by month for 2014. NASA Earth Observatory

The humid subtropical climate zone where winter rainfall (and sometimes snowfall) is associated with large storms that the westerlies steer from west to east. Most summer rainfall occurs during thunderstorms and from occasional tropical cyclones. Humid subtropical climates lie on the east side continents, roughly between latitudes 20° and 40° degrees away from the equator.

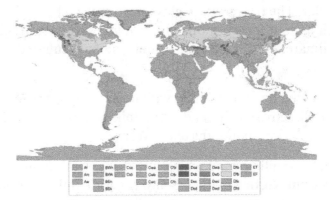

Humid continental climate, worldwide

A humid continental climate is marked by variable weather patterns and a large seasonal temperature variance. Places with more than three months of average daily temperatures above 10 °C (50 °F) and a coldest month temperature below −3 °C (27 °F) and which do not meet the criteria for an arid or semiarid climate, are classified as continental.

An oceanic climate is typically found along the west coasts at the middle latitudes of all the world's continents, and in southeastern Australia, and is accompanied by plentiful precipitation year round.

The Mediterranean climate regime resembles the climate of the lands in the Mediterranean Basin, parts of western North America, parts of Western and South Australia, in southwestern South Africa and in parts of central Chile. The climate is characterized by hot, dry summers and cool, wet winters.

A steppe is a dry grassland with an annual temperature range in the summer of up to 40 °C (104 °F) and during the winter down to −40 °C (−40 °F).

A subarctic climate has little precipitation, and monthly temperatures which are above 10 °C (50 °F) for one to three months of the year, with permafrost in large parts of the area due to the cold winters. Winters within subarctic climates usually include up to six months of temperatures averaging below 0 °C (32 °F).

Tundra occurs in the far Northern Hemisphere, north of the taiga belt, including vast areas of northern Russia and Canada.

A polar ice cap, or polar ice sheet, is a high-latitude region of a planet or moon that is covered in ice. Ice caps form because high-latitude regions receive less energy as solar radiation from the sun than equatorial regions, resulting in lower surface temperatures.

A desert is a landscape form or region that receives very little precipitation. Deserts usually have a large diurnal and seasonal temperature range, with high or low, depending on location daytime temperatures (in summer up to 45 °C or 113 °F), and low nighttime temperatures (in winter down to 0 °C or 32 °F) due to extremely low humidity. Many deserts are formed by rain shadows, as mountains block the path of moisture and precipitation to the desert.

Thornthwaite

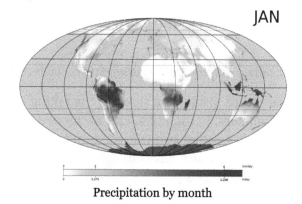

Precipitation by month

Devised by the American climatologist and geographer C. W. Thornthwaite, this climate classification method monitors the soil water budget using evapotranspiration. It monitors the portion of total precipitation used to nourish vegetation over a certain area. It uses indices such as a humidity index and an aridity index to determine an area's moisture regime based upon its average temperature, average rainfall, and average vegetation type. The lower the value of the index in any given area, the drier the area is.

The moisture classification includes climatic classes with descriptors such as hyperhumid, humid, subhumid, subarid, semi-arid (values of −20 to −40), and arid (values below −40). Humid regions experience more precipitation than evaporation each year, while arid regions experience greater evaporation than precipitation on an annual basis. A total of 33 percent of the Earth's land-mass is considered either arid or semi-arid, including southwest North America, southwest South America, most of northern and a small part of southern Africa, southwest and portions of eastern Asia, as well as much of Australia. Studies suggest that precipitation effectiveness (PE) within the Thornthwaite moisture index is overestimated in the summer and underestimated in the winter. This index can be effectively used to determine the number of herbivore and mammal species numbers within a given area. The index is also used in studies of climate change.

Thermal classifications within the Thornthwaite scheme include microthermal, mesothermal, and megathermal regimes. A microthermal climate is one of low annual mean temperatures, generally between 0 °C (32 °F) and 14 °C (57 °F) which experiences short summers and has a potential evaporation between 14 centimetres (5.5 in) and 43 centimetres (17 in). A mesothermal climate lacks persistent heat or persistent cold, with potential evaporation between 57 centimetres (22 in) and 114 centimetres (45 in). A megathermal climate is one with persistent high temperatures and abundant rainfall, with potential annual evaporation in excess of 114 centimetres (45 in).

Record

Modern

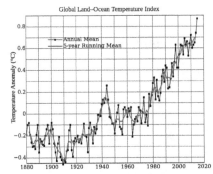

Global mean surface temperature change since 1880. Source: NASA GISS

Details of the modern climate record are known through the taking of measurements from such weather instruments as thermometers, barometers, and anemometers during the past few centuries. The instruments used to study weather over the modern time scale, their known error, their immediate environment, and their exposure have changed over the years, which must be considered when studying the climate of centuries past.

Paleoclimatology

Paleoclimatology is the study of past climate over a great period of the Earth's history. It uses evidence from ice sheets, tree rings, sediments, coral, and rocks to determine the past state of the climate. It demonstrates periods of stability and periods of change and can indicate whether changes follow patterns such as regular cycles.

Climate Change

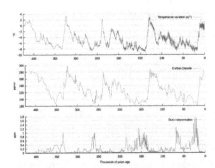

Variations in CO_2, temperature and dust from the Vostok ice core over the past 450,000 years

Climate change is the variation in global or regional climates over time. It reflects changes in the variability or average state of the atmosphere over time scales ranging from decades to millions of years. These changes can be caused by processes internal to the Earth, external forces (e.g. variations in sunlight intensity) or, more recently, human activities.

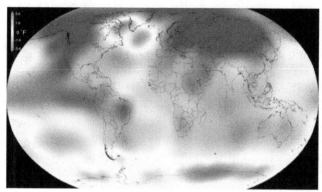

2015 – Warmest Global Year on Record (since 1880) – Colors indicate temperature anomalies
(NASA/NOAA; 20 January 2016).

In recent usage, especially in the context of environmental policy, the term "climate change" often refers only to changes in modern climate, including the rise in average surface temperature known as global warming. In some cases, the term is also used with a presumption of human causation, as in the United Nations Framework Convention on Climate Change (UNFCCC). The UNFCCC uses "climate variability" for non-human caused variations.

Earth has undergone periodic climate shifts in the past, including four major ice ages. These consisting of glacial periods where conditions are colder than normal, separated by interglacial periods. The accumulation of snow and ice during a glacial period increases the surface albedo, reflecting more of the Sun's energy into space and maintaining a lower atmospheric temperature. Increases in greenhouse gases, such as by volcanic activity, can increase the global temperature and produce an interglacial period. Suggested causes of ice age periods include the positions of the continents, variations in the Earth's orbit, changes in the solar output, and volcanism.

Climate Models

Climate models use quantitative methods to simulate the interactions of the atmosphere, oceans, land surface and ice. They are used for a variety of purposes; from the study of the dynamics of the weather and climate system, to projections of future climate. All climate models balance, or very nearly balance, incoming energy as short wave (including visible) electromagnetic radiation to the earth with outgoing energy as long wave (infrared) electromagnetic radiation from the earth. Any imbalance results in a change in the average temperature of the earth.

The most talked-about applications of these models in recent years have been their use to infer the consequences of increasing greenhouse gases in the atmosphere, primarily carbon dioxide. These models predict an upward trend in the global mean surface temperature, with the most rapid increase in temperature being projected for the higher latitudes of the Northern Hemisphere.

Models can range from relatively simple to quite complex:

- Simple radiant heat transfer model that treats the earth as a single point and averages outgoing energy

- this can be expanded vertically (radiative-convective models), or horizontally

- finally, (coupled) atmosphere–ocean–sea ice global climate models discretise and solve the full equations for mass and energy transfer and radiant exchange.

Climate forecasting is a way by some scientists are using to predict climate change. In 1997 the prediction division of the International Research Institute for Climate and Society at Columbia University began generating seasonal climate forecasts on a real-time basis. To produce these forecasts an extensive suite of forecasting tools was developed, including a multimodel ensemble approach that required thorough validation of each model's accuracy level in simulating interannual climate variability.

Weather

Weather is the state of the atmosphere, to the degree that it is hot or cold, wet or dry, calm or stormy, clear or cloudy. Most weather phenomena occur in the troposphere, just below the stratosphere. Weather refers to day-to-day temperature and precipitation activity, whereas climate is the term for the statistics of atmospheric conditions over longer periods of time. When used without qualification, "weather" is generally understood to mean the weather of Earth.

Weather is driven by air pressure, temperature and moisture differences between one place and another. These differences can occur due to the sun's angle at any particular spot, which varies by latitude from the tropics. The strong temperature contrast between polar and tropical air gives rise to the jet stream. Weather systems in the mid-latitudes, such as extratropical cyclones, are caused by instabilities of the jet stream flow. Because the Earth's axis is tilted relative to its orbital plane, sunlight is incident at different angles at different times of the year. On Earth's surface, temperatures usually range ±40 °C (−40 °F to 100 °F) annually. Over thousands of years, changes in Earth's orbit can affect the amount and distribution of solar energy received by the Earth, thus influencing long-term climate and global climate change.

Surface temperature differences in turn cause pressure differences. Higher altitudes are cooler than lower altitudes due to differences in compressional heating. Weather forecasting is the application of science and technology to predict the state of the atmosphere for a future time and a given location. The system is a chaotic system; so small changes to one part of the system can grow to have large effects on the system as a whole. Human attempts to control the weather have occurred throughout human history, and there is evidence that human activities such as agriculture and industry have modified weather patterns.

Studying how the weather works on other planets has been helpful in understanding how weather works on Earth. A famous landmark in the Solar System, Jupiter's *Great Red Spot*, is an anticyclonic storm known to have existed for at least 300 years. However, weather is not limited to planetary bodies. A star's corona is constantly being lost to space, creating what is essentially a very thin atmosphere throughout the Solar System. The movement of mass ejected from the Sun is known as the solar wind.

Causes

Cumulus mediocris cloud surrounded by stratocumulus

On Earth, the common weather phenomena include wind, cloud, rain, snow, fog and dust storms. Less common events include natural disasters such as tornadoes, hurricanes, typhoons and ice storms. Almost all familiar weather phenomena occur in the troposphere (the lower part of the atmosphere). Weather does occur in the stratosphere and can affect weather lower down in the troposphere, but the exact mechanisms are poorly understood.

Weather occurs primarily due to air pressure, temperature and moisture differences between one place to another. These differences can occur due to the sun angle at any particular spot, which varies by latitude from the tropics. In other words, the farther from the tropics one lies, the lower the sun angle is, which causes those locations to be cooler due to the indirect sunlight. The strong temperature contrast between polar and tropical air gives rise to the jet stream. Weather systems in the mid-latitudes, such as extratropical cyclones, are caused by instabilities of the jet stream flow. Weather systems in the tropics, such as monsoons or organized thunderstorm systems, are caused by different processes.

Because the Earth's axis is tilted relative to its orbital plane, sunlight is incident at different angles at different times of the year. In June the Northern Hemisphere is tilted towards the sun, so at any given Northern Hemisphere latitude sunlight falls more directly on that spot than in December. This effect causes seasons. Over thousands to hundreds of thousands of years, changes in Earth's orbital parameters affect the amount and distribution of solar energy received by the Earth and influence long-term climate.

The uneven solar heating (the formation of zones of temperature and moisture gradients, or frontogenesis) can also be due to the weather itself in the form of cloudiness and precipitation. Higher altitudes are typically cooler than lower altitudes, which is explained by the lapse rate. In some situations, the temperature actually increases with height. This phenomenon is known as an inversion and can cause mountaintops to be warmer than the valleys below. Inversions can lead to the formation of fog and often act as a cap that suppresses thunderstorm development. On local scales, temperature differences can occur because different surfaces (such as oceans, forests, ice sheets, or man-made objects) have differing physical characteristics such as reflectivity, roughness, or moisture content.

Surface temperature differences in turn cause pressure differences. A hot surface warms the air above it causing it to expand and lower the density and the resulting surface air pressure. The resulting horizontal pressure gradient moves the air from higher to lower pressure regions, creating a wind, and the Earth's rotation then causes deflection of this air flow due to the Coriolis effect. The simple systems thus formed can then display emergent behaviour to produce more complex systems and thus other weather phenomena. Large scale examples include the Hadley cell while a smaller scale example would be coastal breezes.

The atmosphere is a chaotic system, so small changes to one part of the system can grow to have large effects on the system as a whole. This makes it difficult to accurately predict weather more than a few days in advance, though weather forecasters are continually working to extend this limit through the scientific study of weather, meteorology. It is theoretically impossible to make useful day-to-day predictions more than about two weeks ahead, imposing an upper limit to potential for improved prediction skill.

Shaping The Planet Earth

Weather is one of the fundamental processes that shape the Earth. The process of weathering breaks down the rocks and soils into smaller fragments and then into their constituent substances. During rains precipitation, the water droplets absorb and dissolve carbon dioxide from the surrounding air. This causes the rainwater to be slightly acidic, which aids the erosive properties of water. The released sediment and chemicals are then free to take part in chemical reactions that can affect the surface further (such as acid rain), and sodium and chloride ions (salt) deposited in the seas/oceans. The sediment may reform in time and by geological forces into other rocks and soils. In this way, weather plays a major role in erosion of the surface.

Global Weather Video for Year 2015

EUMETSAT created "A Year in Weather 2015" a narrated video of the earth's weather photographed from weather satellites for the entire year 2015. Geostationary satellite photographs from EUMETSAT, the Japan Meteorological Agency and the National Oceanic and Atmospheric Administration were assembled to show weather changing on earth for 365 days in a time lapse video.

Major Wind and Pressure Systems and Related Weather

Region	Name	Pressure	Surface Winds	Weather
Equator (0°)	Doldrums (ITCZ) (equatorial low)	Low	Light, variable winds	Cloudiness, abundant precipitation in all seasons; breeding ground for hurricanes. Relatively low sea-surface salinity because of high rainfall relative to evaporation
0°–30°N and S	Trade winds (easterlies)		Northeast in Northern Hemisphere; Southeast in Southern Hemisphere	Summer wet, winter dry; pathway for tropical disturbances

30°N and S	Horse latitudes	High	Light, variable winds	Little cloudiness; dry in all seasons. Relatively high sea-surface salinity because of high evaporation relative to precipitation
30°–60°N and S	Prevailing Westerlies	-	Southwest in Northern Hemisphere; Northwest in Southern Hemisphere	Winter wet, summer dry; pathway for subtropical high and low pressure
60°N and S	Polar front	Low	Variable	Stormy, cloudy weather zone; ample precipitation in all seasons
60°–90°N and S	Polar easterlies	-	Northeast in Northern Hemisphere; Southeast in Southern Hemisphere	Cold polar air with very low temperatures
90°N and S	Poles	High	Southerly in Northern Hemisphere; Northerly in Southern Hemisphere	Cold, dry air; sparse precipitation in all seasons

Effect on Humans

Weather, seen from an anthropological perspective, is something all humans in the world constantly experience through their senses, at least while being outside. There are socially and scientifically constructed understandings of what weather is, what makes it change, the effect it has on humans in different situations, etc. Therefore, weather is something people often communicate about.

Effects on Populations

New Orleans, Louisiana, after being struck by Hurricane Katrina. Katrina was a Category 3 hurricane when it struck although it had been a category 5 hurricane in the Gulf of Mexico.

Weather has played a large and sometimes direct part in human history. Aside from climatic changes that have caused the gradual drift of populations (for example the desertification of the Middle East, and the formation of land bridges during glacial periods), extreme weather events have caused smaller scale population movements and intruded directly in historical events. One such event is the saving of Japan from invasion by the Mongol fleet of Kublai Khan by the Kamikaze winds in 1281. French claims to Florida came to an end in 1565 when a hurricane destroyed the French fleet, allowing Spain to conquer Fort Caroline. More recently, Hurricane Katrina redistributed over one million people from the central Gulf coast elsewhere across the United States, becoming the largest diaspora in the history of the United States.

The Little Ice Age caused crop failures and famines in Europe. The 1690s saw the worst famine in

France since the Middle Ages. Finland suffered a severe famine in 1696–1697, during which about one-third of the Finnish population died.

Forecasting

Forecast of surface pressures five days into the future for the north Pacific, North America, and north Atlantic ocean as on 9 June 2008

Weather forecasting is the application of science and technology to predict the state of the atmosphere for a future time and a given location. Human beings have attempted to predict the weather informally for millennia, and formally since at least the nineteenth century. Weather forecasts are made by collecting quantitative data about the current state of the atmosphere and using scientific understanding of atmospheric processes to project how the atmosphere will evolve.

Once an all-human endeavor based mainly upon changes in barometric pressure, current weather conditions, and sky condition, forecast models are now used to determine future conditions. Human input is still required to pick the best possible forecast model to base the forecast upon, which involves pattern recognition skills, teleconnections, knowledge of model performance, and knowledge of model biases. The chaotic nature of the atmosphere, the massive computational power required to solve the equations that describe the atmosphere, error involved in measuring the initial conditions, and an incomplete understanding of atmospheric processes mean that forecasts become less accurate as the difference in current time and the time for which the forecast is being made (the *range* of the forecast) increases. The use of ensembles and model consensus helps to narrow the error and pick the most likely outcome.

There are a variety of end users to weather forecasts. Weather warnings are important forecasts because they are used to protect life and property. Forecasts based on temperature and precipitation are important to agriculture, and therefore to commodity traders within stock markets. Temperature forecasts are used by utility companies to estimate demand over coming days. On an everyday basis, people use weather forecasts to determine what to wear on a given day. Since outdoor activities are severely curtailed by heavy rain, snow and the wind chill, forecasts can be used to plan activities around these events, and to plan ahead and survive them.

Modification

The aspiration to control the weather is evident throughout human history: from ancient rituals intended to bring rain for crops to the U.S. Military Operation Popeye, an attempt to disrupt supply lines by lengthening the North Vietnamese monsoon. The most successful attempts at influencing weather involve cloud seeding; they include the fog- and low stratus dispersion techniques employed by major airports, techniques used to increase winter precipitation over mountains, and techniques to suppress hail. A recent example of weather control was China's preparation for the 2008 Summer Olympic Games. China shot 1,104 rain dispersal rockets from 21 sites in the city of Beijing in an effort to keep rain away from the opening ceremony of the games on 8 August 2008. Guo Hu, head of the Beijing Municipal Meteorological Bureau (BMB), confirmed the success of the operation with 100 millimeters falling in Baoding City of Hebei Province, to the southwest and Beijing's Fangshan District recording a rainfall of 25 millimeters.

Whereas there is inconclusive evidence for these techniques' efficacy, there is extensive evidence that human activity such as agriculture and industry results in inadvertent weather modification:

- Acid rain, caused by industrial emission of sulfur dioxide and nitrogen oxides into the atmosphere, adversely affects freshwater lakes, vegetation, and structures.

- Anthropogenic pollutants reduce air quality and visibility.

- Climate change caused by human activities that emit greenhouse gases into the air is expected to affect the frequency of extreme weather events such as drought, extreme temperatures, flooding, high winds, and severe storms. However, some experts argue these claims are unfounded and take issue with these conclusions.

- Heat, generated by large metropolitan areas have been shown to minutely affect nearby weather, even at distances as far as 1,600 kilometres (990 mi).

The effects of inadvertent weather modification may pose serious threats to many aspects of civilization, including ecosystems, natural resources, food and fiber production, economic development, and human health.

Microscale Meteorology

Microscale meteorology is the study of short-lived atmospheric phenomena smaller than mesoscale, about 1 km or less. These two branches of meteorology are sometimes grouped together as "mesoscale and microscale meteorology" (MMM) and together study all phenomena smaller than synoptic scale; that is they study features generally too small to be depicted on a weather map. These include small and generally fleeting cloud "puffs" and other small cloud features.

Extremes on Earth

On Earth, temperatures usually range ±40 °C (100 °F to −40 °F) annually. The range of climates and latitudes across the planet can offer extremes of temperature outside this range. The coldest air temperature ever recorded on Earth is −89.2 °C (−128.6 °F), at Vostok Station, Antarctica on 21 July 1983. The hottest air temperature ever recorded was 57.7 °C (135.9 °F) at 'Aziziya, Libya, on 13 September 1922, but that reading is queried. The highest recorded average annual temperature was 34.4 °C (93.9 °F) at Dallol, Ethiopia. The coldest recorded average annual temperature was −55.1 °C (−67.2 °F) at Vostok Station, Antarctica.

The coldest average annual temperature in a permanently inhabited location is at Eureka, Nunavut, in Canada, where the annual average temperature is −19.7 °C (−3.5 °F).

Extraterrestrial within the Solar System

Studying how the weather works on other planets has been seen as helpful in understanding how it works on Earth. Weather on other planets follows many of the same physical principles as weather on Earth, but occurs on different scales and in atmospheres having different chemical composition. The Cassini–Huygens mission to Titan discovered clouds formed from methane or ethane which deposit rain composed of liquid methane and other organic compounds. Earth's atmosphere includes six latitudinal circulation zones, three in each hemisphere. In contrast, Jupiter's banded appearance shows many such zones, Titan has a single jet stream near the 50th parallel north latitude, and Venus has a single jet near the equator.

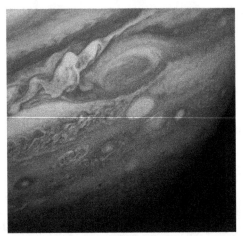

Jupiter's Great Red Spot in 1979

One of the most famous landmarks in the Solar System, Jupiter's *Great Red Spot*, is an anticyclonic storm known to have existed for at least 300 years. On other gas giants, the lack of a surface allows the wind to reach enormous speeds: gusts of up to 600 metres per second (about 2,100 km/h or 1,300 mph) have been measured on the planet Neptune. This has created a puzzle for planetary scientists. The weather is ultimately created by solar energy and the amount of energy received by Neptune is only about of that received by Earth, yet the intensity of weather phenomena on Neptune is far greater than on Earth. The strongest planetary winds discovered so far are on the extrasolar planet HD 189733 b, which is thought to have easterly winds moving at more than 9,600 kilometres per hour (6,000 mph).

Space Weather

Aurora Borealis

Weather is not limited to planetary bodies. Like all stars, the sun's corona is constantly being lost to space, creating what is essentially a very thin atmosphere throughout the Solar System. The movement of mass ejected from the Sun is known as the solar wind. Inconsistencies in this wind and larger events on the surface of the star, such as coronal mass ejections, form a system that has features analogous to conventional weather systems (such as pressure and wind) and is generally known as space weather. Coronal mass ejections have been tracked as far out in the solar system as Saturn. The activity of this system can affect planetary atmospheres and occasionally surfaces.

The interaction of the solar wind with the terrestrial atmosphere can produce spectacular aurorae, and can play havoc with electrically sensitive systems such as electricity grids and radio signals.

Air Mass

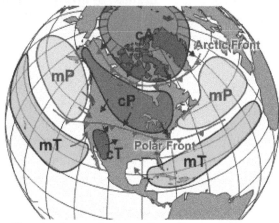

Different air masses which affect North America, as well as other continents, tend to be separated by frontal boundaries

In meteorology, an air mass is a volume of air defined by its temperature and water vapor content. Air masses cover many hundreds or thousands of square miles, and adapt to the characteristics of the surface below them. They are classified according to latitude and their continental or maritime source regions. Colder air masses are termed polar or arctic, while warmer air masses are deemed tropical. Continental and superior air masses are dry while maritime and monsoon air masses are moist. Weather fronts separate air masses with different density (temperature and/or moisture) characteristics. Once an air mass moves away from its source region, underlying vegetation and water bodies can quickly modify its character. Classification schemes tackle an air mass' character-istics, and well as modification.

Classification and Notation

The Bergeron classification is the most widely accepted form of air mass classification, though others have produced more refined versions of this scheme over different regions of the globe. Air mass classification involves three letters. The first letter describes its moisture properties, with c used for continental air masses (dry) and m for maritime air masses (moist). The second letter describes the thermal characteristic of its source region: T for Tropical, P for Polar, A for arctic or Antarctic, M for monsoon, E for Equatorial, and S for superior air (an adiabatically drying and warming air formed by significant downward motion in the atmosphere). For instance, an air mass originating over the desert southwest of the United States in summer may be designated "cT". An air mass originating over northern Siberia in winter may be indicated as "cA".

The stability of an air mass may be shown using a third letter, either "k" (air mass colder than the surface below it) or "w" (air mass warmer than the surface below it). An example of this might be a polar air mass blowing over the Gulf Stream, denor it is replacing is colder than the replaced air mass (usually for polar air masses). For example, a series of fronts over the Pacific might show an air mass denoted mPk followed by another denoted mPk'.

Another convention utilizing these symbols is the indication of modification or transformation of one type to another. For instance, an Arctic air mass blowing out over the Gulf of Alaska may be shown as "cA-mPk". Yet another convention indicates the layering of air masses in certain situations. For instance, the overrunning of a polar air mass by an air mass from the Gulf of Mexico over the Central United States might be shown with the notation "mT/cP" (sometimes using a horizontal line as in fraction notation).

Characteristics

Arctic, Antarctic, and polar air masses are cold. The qualities of arctic air are developed over ice and snow-covered ground. Arctic air is deeply cold, colder than polar air masses. Arctic air can be shallow in the summer, and rapidly modify as it moves equatorward. Polar air masses develop over higher latitudes over the land or ocean, are very stable, and generally shallower than arctic air. Polar air over the ocean (maritime) loses its stability as it gains moisture over warmer ocean waters.

Tropical and equatorial air masses are hot as they develop over lower latitudes. Those that develop over land (continental) are drier and hotter than those that develop over oceans, and travel poleward on the western periphery of the subtropical ridge. Maritime tropical air masses are sometimes referred to as trade air masses. Monsoon air masses are moist and unstable. Superior air masses are dry, and rarely reach the ground. They normally reside over maritime tropical air masses, forming a warmer and drier layer over the more moderate moist air mass below, forming what is known as a trade wind inversion over the maritime tropical air mass. Continental Polar air masses (cP) are air masses that are cold and dry due to their continental source region. Continental polar air masses that affect North America form over interior Canada. Continental Tropical air masses (cT) are a type of tropical air produced by the subtropical ridge over large areas of land and typically originate from low-latitude deserts such as the Sahara Desert in northern Africa, which is the major source of these air masses. Other less important sources producing cT air masses are the Arabian Peninsula, the central arid/semi-arid part of Australia and deserts lying in the Southwestern United States. Continental tropical air masses are extremely hot and dry.

Movement and Fronts

A weather front is a boundary separating two masses of air of different densities, and is the principal cause of meteorological phenomena. In surface weather analyses, fronts are depicted using various colored lines and symbols, depending on the type of front. The air masses separated by a front usually differ in temperature and humidity. Cold fronts may feature narrow bands of thunderstorms and severe weather, and may on occasion be preceded by squall lines or dry lines. Warm fronts are usually preceded by stratiform precipitation and fog. The weather usually clears quickly after a front's passage. Some fronts produce no precipitation and little cloudiness, although there is invariably a wind shift.

Cold fronts and occluded fronts generally move from west to east, while warm fronts move poleward. Because of the greater density of air in their wake, cold fronts and cold occlusions move faster than warm fronts and warm occlusions. Mountains and warm bodies of water can slow the movement of fronts. When a front becomes stationary, and the density contrast across the frontal boundary vanishes, the front can degenerate into a line which separates regions of differing wind velocity, known as a shearline. This is most common over the open ocean.

Modification

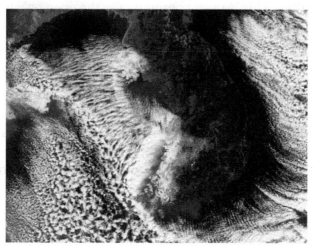

Lake-effect snow bands near the Korean Peninsula

Air masses can be modified in a variety of ways. Surface flux from underlying vegetation, such as forest, acts to moisten the overlying air mass. Heat from underlying warmer waters can significantly modify an air mass over distances as short as 35 kilometres (22 mi) to 40 kilometres (25 mi). For example, southwest of extratropical cyclones, curved cyclonic flow bringing cold air across the relatively warm water bodies can lead to narrow lake-effect snow bands. Those bands bring strong localized precipitation since large water bodies such as lakes efficiently store heat that results in significant temperature differences (larger than 13°C or 23°F) between the water surface and the air above. Because of this temperature difference, warmth and moisture are transported upward, condensing into vertically oriented clouds which produce snow showers. The temperature decrease with height and cloud depth are directly affected by both the water temperature and the large-scale environment. The stronger the temperature decrease with height, the deeper the clouds get, and the greater the precipitation rate becomes.

Weather Front

A weather front is a boundary separating two masses of air of different densities, and is the principal cause of meteorological phenomena. In surface weather analyses, fronts are depicted using various colored triangles and half-circles, depending on the type of front. The air masses separated by a front usually differ in temperature and humidity. Cold fronts may feature narrow bands of thunderstorms and severe weather, and may on occasion be preceded by squall lines or dry lines. Warm fronts are usually preceded by stratiform precipitation and fog. The weather usually clears quickly after a front's passage. Some fronts produce no precipitation and little cloudiness, although there is invariably a wind shift.

Cold fronts and occluded fronts generally move from west to east, while warm fronts move poleward. Because of the greater density of air in their wake, cold fronts and cold occlusions move faster than warm fronts and warm occlusions. Mountains and warm bodies of water can slow the movement of fronts. When a front becomes stationary, and the density contrast across the frontal boundary vanishes, the front can degenerate into a line which separates regions of differing wind velocity, known as a shearline. This is most common over the open ocean.

Bergeron Classification of Air Masses

The Bergeron classification is the most widely accepted form of air mass classification. Air mass classification involves three letters. The first letter describes its moisture properties, with c used for continental air masses (dry) and m for maritime air masses (moist). The second letter describes the thermal characteristic of its source region: T for tropical, P for polar, A for arctic or Antarctic, M for monsoon, E for equatorial, and S for superior air (dry air formed by significant upward motion in the atmosphere). The third letter is used to designate the stability of the atmosphere. If the air mass is colder than the ground below it, it is labeled k. If the air mass is warmer than the ground below it, it is labeled w. Fronts separate air masses of different types or origins, and are located along troughs of lower pressure.

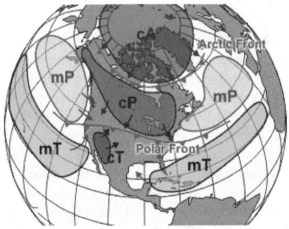

Different air masses which affect North America, as well as other continents, tend to be separated by frontal boundaries. In this illustration, the Arctic front separates Arctic from Polar air masses, while the Polar front separates Polar air from warm air masses. (cA is continental arctic; cP is continental polar; mP is maritime polar; cT is continental tropic; and mT is maritime tropic.)

Surface Weather Analysis

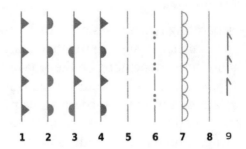

A guide to the symbols for weather fronts that may be found on a weather map:

1. cold front

2. warm front

3. stationary front

4. occluded front

5. surface trough

6. squall/shear line

7. dry line

8. tropical wave

9. trowal

A surface weather analysis is a special type of weather map which provides a view of weather elements over a geographical area at a specified time based on information from ground-based weather stations. Weather maps are created by plotting or tracing the values of relevant quantities such as sea-level pressure, temperature, and cloud cover onto a geographical map to help find synoptic scale features such as weather fronts. Surface weather analyses have special symbols which show frontal systems, cloud cover, precipitation, or other important information. For example, an *H* may represent high pressure, implying fair weather. An *L* on the other hand may represent low pressure, which frequently accompanies precipitation. Low pressure also creates surface winds deriving from high pressure zones. Various symbols are used not just for frontal zones and other surface boundaries on weather maps, but also to depict the present weather at various locations on the weather map. In addition, areas of precipitation help determine the frontal type and location.

Front Types

There are two different words used within meteorology to describe weather around a frontal zone. The term "anafront" describes boundaries which show instability, meaning air rises rapidly along and over the boundary to cause significant weather changes. A "katafront" is weaker, bringing smaller changes in temperature and moisture, as well as limited rainfall.

Cold Front

A cold front is located at the leading edge of the temperature drop off, which in an isotherm analysis shows up as the leading edge of the isotherm gradient, and it normally lies within a sharp surface trough. Cold fronts often bring heavy thunderstorms, rain and hail. Cold front can produce sharper changes in weather and move up to twice as quickly as warm fronts, since cold air is denser than warm air and rapidly replaces the warm air preceding the boundary. On weather maps, the surface position of the cold front is marked with the symbol of a blue line of triangle-shaped pips pointing in the direction of travel, and it is placed at the leading edge of the cooler air mass. Cold fronts come in association with a low-pressure area. The concept of colder, dense air "wedging" under the less dense warmer air is often used to depict how air is lifted along a frontal boundary. The cold air wedging underneath warmer air creates the strongest winds just above the ground surface, a phenomenon often associated with property-damaging wind gusts. This lift would then form a narrow line of showers and thunderstorms if enough moisture were present. However, this concept isn't an accurate description of the physical processes; upward motion is not produced because of warm air "ramping up" cold, dense air, rather, frontogenetical circulation is behind the upward forcing.

Warm Front

Warm fronts are at the leading edge of a homogeneous warm air mass, which is located on the equatorward edge of the gradient in isotherms, and lie within broader troughs of low pressure than cold fronts. A warm front moves more slowly than the cold front which usually follows because cold air is denser and harder to remove from the Earth's surface. This also forces temperature differences across warm fronts to be broader in scale. Clouds ahead of the warm front are mostly stratiform, and rainfall gradually increases as the front approaches. Fog can also occur preceding a warm frontal passage. Clearing and warming is usually rapid after frontal passage. If the warm air mass is unstable, thunderstorms may be embedded among the stratiform clouds ahead of the front, and after frontal passage thundershowers may continue. On weather maps, the surface location of a warm front is marked with a red line of semicircles pointing in the direction of travel.

Occluded Front

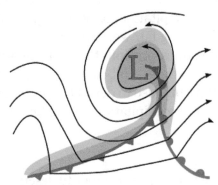

Occluded front depiction for the Northern Hemisphere

An occluded front is formed when a cold front overtakes a warm front. and usually form around mature low-pressure areas. The cold and warm fronts curve naturally poleward into the point of occlusion, which is also known as the triple point. It lies within a sharp trough, but the air mass behind the boundary can be either warm or cold. In a cold occlusion, the air mass overtaking the warm front is cooler than the cool air ahead of the warm front and plows under both air masses. In a warm occlusion, the air mass overtaking the warm front is warmer than the cold air ahead of the warm front and rides over the colder air mass while lifting the warm air.

A wide variety of weather can be found along an occluded front, with thunderstorms possible, but usually their passage is associated with a drying of the air mass. Within the occlusion of the front, a circulation of air brings warm air upward and sends drafts of cold air downward, or vice-versa depending on the occlusion the front is experiencing. Precipitations and clouds are associated with the *trowal*, the projection on the Earth's surface of the tongue of warm air aloft formed during the occlusion process of the depression.

Occluded fronts are indicated on a weather map by a purple line with alternating half-circles and triangles pointing in direction of travel. The trowal is indicated by a series of blue and red junction lines.

Stationary Front

A stationary front is a non-moving (or stalled) boundary between two air masses, neither of which is strong enough to replace the other. They tend to remain essentially in the same area for extended periods of time, usually moving in waves. There is normally a broad temperature gradient behind the boundary with more widely spaced isotherm packing.

A wide variety of weather can be found along a stationary front, but usually clouds and prolonged precipitation are found there. Stationary fronts either dissipate after several days or devolve into shear lines, but they can transform into a cold or warm front if conditions aloft change. Stationary fronts are marked on weather maps with alternating red half-circles and blue spikes pointing in opposite directions, indicating no significant movement.

When stationary fronts become smaller in scale, degenerating to a narrow zone where wind direction changes significantly over a relatively short distance, they become known as shearlines. A shearline is depicted as a line of red dots and dashes.

Dry Line

A similar phenomenon to a weather front is the dry line, which is the boundary between air masses with significant moisture differences. When the westerlies increase on the north side of surface highs, areas of lowered pressure will form downwind of north–south oriented mountain chains, leading to the formation of a lee trough. Near the surface during daylight hours, warm moist air is denser than dry air of greater temperature, and thus the warm moist air wedges under the drier air like a cold front. At higher altitudes, the warm moist air is less dense than the dry air and the boundary slope reverses. In the vicinity of the reversal aloft, severe weather is possible, especially when a triple point is formed with a cold front. A weaker form of the dry line seen more commonly is the lee trough, which displays weaker differences in moisture. When moisture pools along the boundary during the warm season, it can be the focus of diurnal thunderstorms.

The dry line may occur anywhere on earth in regions intermediate between desert areas and warm seas. The southern plains west of the Mississippi River in the United States are a particularly favored location. The dry line normally moves eastward during the day and westward at night. A dry line is depicted on National Weather Service (NWS) surface analyses as an orange line with scallops facing into the moist sector. Dry lines are one of the few surface fronts where the pips indicated do not necessarily reflect the direction of motion.

Squall Line

Organized areas of thunderstorm activity not only reinforce pre-existing frontal zones, but can outrun cold fronts in a pattern where the upper level jet splits apart into two streams, with the resultant Mesoscale Convective System (MCS) forming at the point of the upper level split in the wind pattern running southeast into the warm sector parallel to low-level thickness lines. When the convection is strong and linear or curved, the MCS is called a squall line, with the feature placed at the leading edge of the significant wind shift and pressure rise. Even weaker and less organized areas of thunderstorms lead to locally cooler air and higher pressures, and outflow boundaries exist ahead of this type of activity, which can act as foci for additional thunderstorm activity later in the day.

A shelf cloud such as this one can be a sign that a squall is imminent

These features are often depicted in the warm season across the United States on surface analyses and lie within surface troughs. If outflow boundaries or squall lines form over arid regions, a haboob may result. Squall lines are depicted on NWS surface analyses as an alternating pattern of two red dots and a dash labelled SQLN or SQUALL LINE, while outflow boundaries are depicted as troughs with a label of OUTFLOW BOUNDARY.

Precipitation Produced

Convective precipitation

Fronts are the principal cause of significant weather. *Convective precipitation* (showers, thundershowers, and related unstable weather) is caused by air being lifted and condensing into clouds by the movement of the cold front or cold occlusion under a mass of warmer, moist air. If the temperature differences of the two air masses involved are large and the turbulence is extreme because of wind shear and the presence of a strong jet stream, "roll clouds" and tornadoes may occur.

In the warm season, lee troughs, breezes, outflow boundaries and occlusions can lead to convection if enough moisture is available. *Orographic precipitation* is precipitation created through the lifting action of air moving over terrain such as mountains and hills, which is most common behind cold fronts that move into mountainous areas. It may sometimes occur in advance of warm fronts moving northward to the east of mountainous terrain. However, precipitation along warm fronts is relatively steady, as in rain or drizzle. Fog, sometimes extensive and dense, often occurs in pre-warm-frontal areas. Although, not all fronts produce precipitation or even clouds because moisture must be present in the air mass which is being lifted.

Movement

Fronts are generally guided by winds aloft, but do not move as quickly. Cold fronts and occluded fronts in the Northern Hemisphere usually travel from the northwest to southeast, while warm fronts move more poleward with time. In the Northern Hemisphere a warm front moves from southwest to northeast. In the Southern Hemisphere, the reverse is true; a cold front usually moves from southwest to northeast, and a warm front moves from northwest to southeast. Movement is largely caused by the pressure gradient force (horizontal differences in atmospheric pressure) and the Coriolis effect, which is caused by Earth's spinning about its axis. Frontal zones can be slowed down by geographic features like mountains and large bodies of warm water.

Atmospheric Pressure

Atmospheric pressure, sometimes also called barometric pressure, is the pressure exerted by the weight of air in the atmosphere of Earth (or that of another planet). In most circumstances atmospheric pressure is closely approximated by the hydrostatic pressure caused by the weight of air above the measurement point. Low-pressure areas have less atmospheric mass above their location, whereas high-pressure areas have more atmospheric mass above their location. Likewise, as elevation increases, there is less overlying atmospheric mass, so that atmospheric pressure decreases with increasing elevation. On average, a column of air one square centimetre [cm^2] (0.16 sq in) in cross-section, measured from sea level to the top of the atmosphere, has a mass of about 1.03 kilograms (2.3 lb) and weight of about 10.1 newtons (2.3 lb_f). That force (across one square centimeter) is a pressure of 10.1 N/cm^2 or 101,000 N/m^2. A column 1 square inch (6.5 cm^2) in cross-section would have a weight of about 14.7 lb (6.7 kg) or about 65.4 N.

Standard Atmospheric

The standard atmosphere (symbol: atm) is a unit of pressure defined as 101325 Pa (1.01325 bar), equivalent to 760 mmHg (torr), 29.92 inHg and 14.696 psi.

Mean Sea Level Pressure

The mean sea level pressure (MSLP) is the atmospheric pressure at sea level. This is the atmospheric pressure normally given in weather reports on radio, television, and newspapers or on the Internet. When barometers in the home are set to match the local weather reports, they measure pressure adjusted to sea level, not the actual local atmospheric pressure.

The *altimeter setting* in aviation, is an atmospheric pressure adjustment.

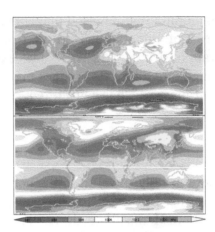

15 year average mean sea level pressure for June, July, and August (top) and December, January, and February (bottom). ERA-15 re-analysis.

Kollsman-type barometric aircraft altimeter (as used in North America) displaying an altitude of 80 ft (24 m).

Average *sea-level pressure* is 1013.25 mbar (101.325 kPa; 29.921 inHg; 760.00 mmHg). In aviation weather reports (METAR), QNH is transmitted around the world in millibars or hectopascals (1 hectopascal = 1 millibar), except in the United States, Canada, and Colombia where it is reported in inches (to two decimal places) of mercury. The United States and Canada also report *sea level pressure* SLP, which is adjusted to sea level by a different method, in the remarks section, not in the internationally transmitted part of the code, in hectopascals or millibars. However, in Canada's public weather reports, sea level pressure is instead reported in kilopascals.

In the US weather code remarks, three digits are all that are transmitted; decimal points and the one or two most significant digits are omitted: 1013.2 mbar (101.32 kPa) is transmitted as 132; 1000.0 mbar (100.00 kPa) is transmitted as 000; 998.7 mbar is transmitted as 987; etc. The highest *sea-level pressure* on Earth occurs in Siberia, where the Siberian High often attains a *sea-level pressure* above 1050 mbar (105 kPa; 31 inHg), with record highs close to 1085 mbar (108.5 kPa; 32.0 inHg). The lowest measurable *sea-level pressure* is found at the centers of tropical cyclones and tornadoes, with a record low of 870 mbar (87 kPa; 26 inHg).

Altitude Variation

Pressure varies smoothly from the Earth's surface to the top of the mesosphere. Although the pressure changes with the weather, NASA has averaged the conditions for all parts of the earth year-round. As altitude increases, atmospheric pressure decreases. One can calculate the atmospheric pressure at a given altitude. Temperature and humidity also affect the atmospheric pressure, and it is necessary to know these to compute an accurate figure. The graph at right was developed for a temperature of 15 °C and a relative humidity of 0%.

A very local storm above Snæfellsjökull, showing clouds formed on the mountain by orographic lift

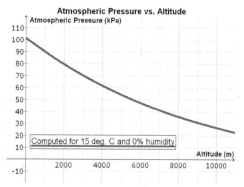

Variation in atmospheric pressure with altitude, computed for 15 °C and 0% relative humidity.

This plastic bottle was sealed at approximately 14,000 feet (4,300 m) altitude, and was crushed by the increase in atmospheric pressure —at 9,000 feet (2,700 m) and 1,000 feet (300 m)— as it was brought down towards sea level.

At low altitudes above the sea level, the pressure decreases by about 1.2 kPa for every 100 meters. For higher altitudes within the troposphere, the following equation (the barometric formula) relates atmospheric pressure p to altitude h

$$p = p_0 \cdot \left(1 - \frac{L \cdot h}{T_0}\right)^{\frac{g \cdot M}{R_0 \cdot L}} \approx p_0 \cdot \left(1 - \frac{g \cdot h}{c_p \cdot T_0}\right)^{\frac{c_p \cdot M}{R_0}},$$

$$p \approx p_0 \cdot \exp\left(-\frac{g \cdot M \cdot h}{R_0 \cdot T_0} \right)$$

where the constant parameters are as described below:

Parameter	Description	Value
p_0	sea level standard atmospheric pressure	101325 Pa
L	temperature lapse rate, $= g/c_p$ for dry air	0.0065 K/m
c_p	constant pressure specific heat	~ 1007 J/(kg•K)
T_0	sea level standard temperature	288.15 K
g	Earth-surface gravitational acceleration	9.80665 m/s²
M	molar mass of dry air	0.0289644 kg/mol
R_0	universal gas constant	8.31447 J/(mol•K)

Local Variation

Hurricane Wilma on 19 October 2005; 882 hPa (12.79 psi) in the storm's eye

Atmospheric pressure varies widely on Earth, and these changes are important in studying weather and climate. See pressure system for the effects of air pressure variations on weather.

Atmospheric pressure shows a diurnal or semidiurnal (twice-daily) cycle caused by global atmospheric tides. This effect is strongest in tropical zones, with an amplitude of a few millibars, and almost zero in polar areas. These variations have two superimposed cycles, a circadian (24 h) cycle and semi-circadian (12 h) cycle.

Records

The highest adjusted-to-sea level barometric pressure ever recorded on Earth (above 750 meters) was 1,085.7 hectopascals (32.06 inHg) measured in Tosontsengel, Mongolia on 19 December 2001. The highest adjusted-to-sea level barometric pressure ever recorded (below 750 meters) was at Agata, Evenkiyskiy, Russia [66°53'N, 93°28'E, elevation: 261 m (856.3 ft)] on 31 December 1968 of 1,083.8 hectopascals (32.00 inHg). The discrimination is due to the problematic assumptions (assuming a standard lapse rate) associated with reduction of sea level from high elevations.

The Dead Sea, the lowest place on Earth at 425 metres (1400 feet) below sea level, has a correspondingly high typical atmospheric pressure of 1065 hPa.

The lowest non-tornadic atmospheric pressure ever measured was 0.858 atm (25.69 inHg), 870 hPa, set on 12 October 1979, during Typhoon Tip in the western Pacific Ocean. The measurement was based on an instrumental observation made from a reconnaissance aircraft.

Measurement Based on Depth of Water

One atmosphere (101 kPa or 14.7 psi) is the pressure caused by the weight of a column of fresh water of approximately 10.3 m (33.8 ft). Thus, a diver 10.3 m underwater experiences a pressure of about 2 atmospheres (1 atm of air plus 1 atm of water). Conversely, 10.3 m is the maximum height to which water can be raised using suction under standard atmospheric conditions.

Low pressures such as natural gas lines are sometimes specified in inches of water, typically written as *w.c.* (water column) or *w.g.* (inches water gauge). A typical gas-using residential appliance in the US is rated for a maximum of 14 w.c., which is approximately 35 hPa. Similar metric units with a wide variety of names and notation based on millimetres, centimetres or metres are now less commonly used.

Boiling Point of Water

Boiling water

Clean fresh water boils at about 100 °C (212 °F) at earth's standard atmospheric pressure. The boiling point is the temperature at which the vapor pressure is equal to the atmospheric pressure around the water. Because of this, the boiling point of water is lower at lower pressure and higher at higher pressure. This is why cooking at high elevations requires adjustments to recipes. A rough approximation of elevation can be obtained by measuring the temperature at which water boils; in the mid-19th century, this method was used by explorers.

Measurement and Maps

An important application of the knowledge that atmospheric pressure varies directly with altitude was in determining the height of hills and mountains thanks to the availability of reliable pressure measurement devices. While in 1774 Maskelyne was confirming Newton's theory of gravitation at and on Schiehallion in Scotland (using plumb bob deviation to show the effect of "gravity") and accurately measure elevation, William Roy using barometric pressure was able to confirm his height determinations, the agreement being to within one meter (3.28 feet). This was then a useful tool for survey work and map making and long has continued to be useful. It was part of the "application of science" which gave practical people the insight that applied science could easily and relatively cheaply be "useful".

Dew Point

In simple terms, the dew point (dew point temperature or dewpoint) is the temperature at which a given concentration of water vapor in air will form dew. More specifically it is a measure of atmospheric moisture. It is the temperature to which air must be cooled at constant pressure and water content to reach saturation. A higher dew point indicates more moisture in the air; a dew point greater than 20 °C (68 °F) is considered uncomfortable and greater than 22 °C (72 °F) is considered to be extremely humid. Frost point is the dew point when temperatures are below freezing.

Humidity

Other things being equal, as the temperature falls, the relative humidity rises, reaching 100% at the dew point, at least at ground level. Dew point temperature is never greater than the air temperature, since the relative humidity cannot exceed 100%.

A technical definition follows: The dew point is the temperature at which the water vapor in a sample of air at constant barometric pressure condenses into liquid water at the same rate at which it evaporates. At temperatures below the dew point, the rate of condensation will be greater than that of evaporation, forming more liquid water. The condensed water is called dew when it forms on a solid surface. The condensed water is called either fog or a cloud, depending on its altitude, when it forms in the air.

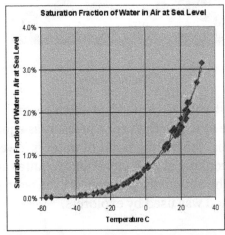

This graph shows the maximum percentage, by mass, of water vapor that air at sea-level pressure across a range of temperatures can contain. For a lower ambient pressure, a curve has to be drawn above the current curve. A higher ambient pressure yields a curve under the current curve.

The dew point is the saturation temperature for water in air. The dew point is associated with relative humidity. A high relative humidity implies that the dew point is closer to the current air temperature. Relative humidity of 100% indicates the dew point is equal to the current temperature and that the air is maximally saturated with water. When the moisture content remains constant and temperature increases, relative humidity decreases.

General aviation pilots use dew point data to calculate the likelihood of carburetor icing and fog, and to estimate the height of a cumuliform cloud base.

At a given temperature but *independent* of barometric pressure, the dew point is a consequence of the absolute humidity, the mass of water per unit volume of air. If both the temperature and pressure rise, however, the dew point will increase and the relative humidity will decrease accordingly. Reducing the absolute humidity without changing other variables will bring the dew point back down to its initial value. In the same way, increasing the absolute humidity after a temperature drop brings the dew point back down to its initial level. If the temperature rises in conditions of constant pressure, then the dew point will remain constant but the relative humidity will drop. For this reason, a constant relative humidity with different temperatures implies that when it is hotter, a higher fraction of the air is present as water vapor, as compared to when it is cooler.

At a given barometric pressure but *independent* of temperature, the dew point indicates the mole fraction of water vapor in the air, or, put differently, determines the specific humidity of the air. If the pressure rises without changing this mole fraction, the dew point will rise accordingly; Reducing the mole fraction, i.e., making the air less humid, would bring the dew point back down to its initial value. In the same way, increasing the mole fraction after a pressure drop brings the relative humidity back up to its initial level. Considering New York (33 ft (10 m) elevation) and Denver (5,280 ft (1,610 m)) elevation), for example, this means that if the dew point and temperature in both cities are the same, then the mass of water vapor per cubic meter of air will be the same, but the mole fraction of water vapor in the air will be greater in Denver.

Relationship to Human Comfort

When the air temperature is high, the human body uses the evaporation of perspiration to cool down, with the cooling effect directly related to how fast the perspiration evaporates. The rate at which perspiration can evaporate depends on how much moisture is in the air and how much moisture the air can hold. If the air is already saturated with moisture, perspiration will not evaporate. The body's thermoregulation cooling system will produce perspiration in an effort to keep the body at its normal temperature even when the rate it is producing sweat exceeds the evaporation rate. So even without generating additional body heat by exercising, one can become coated with sweat on humid days.

As the air surrounding one's body is warmed by body heat, it will rise and be replaced with other air. If air is moved away from one's body with a natural breeze or a fan, sweat will evaporate faster, making perspiration more effective at cooling the body. The more unevaporated perspiration, the greater the discomfort.

A wet bulb thermometer also uses evaporative cooling, so it provides a good measure for use in evaluating comfort level.

Discomfort also exists when the dew point is low (below around –30 °C (–22 °F)). The drier air can cause skin to crack and become irritated more easily. It will also dry out the respiratory paths. OSHA recommends indoor air be maintained at 20 to 24.5 °C (68.0 to 76.1 °F) with a 20-60% relative humidity (a dew point of –4.5 to 15.5 °C (23.9 to 59.9 °F)).

Lower dew points, less than 10 °C (50 °F), correlate with lower ambient temperatures and the body requires less cooling. A lower dew point can go along with a high temperature only at extremely low relative humidity, allowing for relatively effective cooling.

There is some acclimation to higher dew points by those who inhabit tropical and subtropical climates. Thus, a resident of Darwin or Miami, for example, might have a higher threshold for discomfort than a resident of a temperate climate like London or Chicago. Those accustomed to temperate climates often begin to feel uncomfortable when the dew point reaches between 15 and 20 °C (59 and 68 °F)...while others might find dew points below 18°C(65 F) comfortable. Most inhabitants of these areas will consider dew points above 21 °C or 70°F oppressive and tropical-like.

Dew point		Human perception	Relative humidity at 32 °C (90 °F)
Over 26 °C	Over 80 °F	Severely high. Even deadly for asthma related illnesses	73% and higher
24–26 °C	75–80 °F	Extremely uncomfortable, fairly oppressive	62–72%
21–24 °C	70–74 °F	Very humid, quite uncomfortable	52–61%
18–21 °C	65–69 °F	Somewhat uncomfortable for most people at upper edge	44–51%
16–18 °C	60–64 °F	OK for most, but all perceive the humidity at upper edge	37–43%
13–16 °C	55–59 °F	Comfortable	31–36%
10–12 °C	50–54 °F	Very comfortable	26–30%
Under 10 °C	Under 50 °F	A bit dry for some	25% and lower

Measurement

Devices called hygrometers are used to measure dew point over a wide range of temperatures. These devices consist of a polished metal mirror which is cooled as air is passed over it. The temperature at which dew forms is, by definition, the dew point. Manual devices of this sort can be used to calibrate other types of humidity sensors, and automatic sensors may be used in a control loop with a humidifier or dehumidifier to control the dew point of the air in a building or in a smaller space for a manufacturing process.

Extreme Values

A dew point of 33 °C (91 °F) was observed at 2:00 p.m. on July 12, 1987, in Melbourne, Florida. A dew point of 32 °C (90 °F) has been observed in the United States on at least two occasions: Appleton, Wisconsin, at 5:00 p.m. on July 13, 1995, and New Orleans Naval Air Station at 5:00 p.m. on July 30, 1987. A dew point of 35 °C (95 °F) was observed at Dhahran, Saudi Arabia, at 3:00 p.m. on July 8, 2003. Dew points this high are extremely rare occurrences.

Calculating the Dew Point

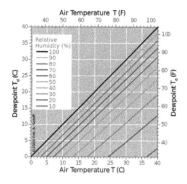

A well-known approximation used to calculate the dew point, T_{dp}, given just the actual ("dry bulb") air temperature, T (in degrees Celsius) and relative humidity (in percent), RH, is the *Magnus formula*:

$$\gamma(T, RH) = \ln\left(\frac{RH}{100}\right) + \frac{bT}{c+T};$$

$$T_{dp} = \frac{c\gamma(T, RH)}{b - \gamma(T, RH)};$$

The more complete formulation and origin of this approximation involves the interrelated saturated water vapor pressure (in units of millibar, which is also hPa) at T, $P_s(T)$, and the actual vapor pressure (also in units of millibar), $P_a(T)$, which can be either found with RH or approximated with the barometric pressure (in millibar units), BP_{mb}, and "wet-bulb" temperature, T_w is:

Note: unless declared otherwise, all temperatures are expressed in degrees Celsius

$$P_s(T) = \frac{100}{RH}P_a(T) = a\exp\left(\frac{bT}{c+T}\right);$$

$$P_a(T) = \frac{RH}{100}P_s(T) = a\exp(\gamma(T, RH)),$$

$$\approx P_s(T_w) - BP_{mb}\,0.00066\left[1 + (0.00115T_w)\right](T - T_w);$$

$$T_{dp} = \frac{c\ln(P_a(T)/a)}{b - \ln(P_a(T)/a)};$$

For greater accuracy, $P_s(T)$ (and, therefore, $\gamma(T,RH)$) can be enhanced, using part of the *Bögel modification*, also known as the Arden Buck equation, which adds a fourth constant d:

$$P_{s:m}(T) = a\exp\left(\left(b - \frac{T}{d}\right)\left(\frac{T}{c+T}\right)\right);$$

$$\gamma_m(T, RH) = \ln\left(\frac{RH}{100}\exp\left(\left(b - \frac{T}{d}\right)\left(\frac{T}{c+T}\right)\right)\right);$$

$$T_{dp} = \frac{c\gamma_m(T, RH)}{b - \gamma_m(T, RH)};$$

(where $a = 6.1121$ millibar; $b = 18.678$; $c = 257.14°C$; $d = 234.5°C$.)

There are several different constant sets in use. The ones used in NOAA's presentation are taken from a 1980 paper by David Bolton in the *Monthly Weather Review*:

$$a\ \ = 6.112\ \text{millibar};\quad b\ \ = 17.67;\quad c\ \ = 243.5°\text{C};$$

These valuations provide a maximum error of 0.1%, for

-30°C ≤ T ≤ +35°C;

1% < RH < 100%;

Also noteworthy is the Sonntag1990,

$$a = 6.112\ \text{millibar};\quad b = 17.62;\quad c = 243.12°\text{C}:\ \ -45°\text{C} \le T \le +60°\text{C}\quad (\pm 0.35°\text{C})$$

Another common set of values originates from the 1974 *Psychrometry and Psychrometric Charts*, as presented by *Paroscientific*,

$$a = 6.105\ \text{millibar};\quad b = 17.27;\quad c = 237.7°\text{C}:\ \ 0°\text{C} \le T \le +60°\text{C}\quad (\pm 0.4°\text{C})$$

Also, in the *Journal of Applied Meteorology and Climatology*, Arden Buck presents several different valuation sets, with different minimum accuracies for different temperature ranges. Two particular sets provide a range of -40 °C +50 °C between the two, with even greater minimum accuracy than all of the other, above sets (maximum error at given |C°| extreme):

$$a = 6.1121\ \text{millibar};\quad b = 17.368;\quad c = 238.88°\text{C}:\ \ \ \ 0°\text{C} \le T \le +50°\text{C}\quad (\le 0.05\%)$$

$$a = 6.1121\ \text{millibar};\quad b = 17.966;\quad c = 247.15°\text{C}:\ \ -40°\text{C} \le T \le 0°\text{C}\quad (\le 0.06\%)$$

Simple Approximation

There is also a very simple approximation that allows conversion between the dew point, temperature and relative humidity. This approach is accurate to within about ±1 °C as long as the relative humidity is above 50%:

$$T_{dp} \approx T - \frac{100 - RH}{5};$$

and

$$RH \approx 100 - 5(T - T_{dp});$$

This can be expressed as a simple rule of thumb:

For every 1°C difference in the dew point and dry bulb temperatures, the relative humidity decreases by 5%, starting with RH = 100% when the dew point equals the dry bulb temperature.

The derivation of this approach, a discussion of its accuracy, comparisons to other approximations, and more information on the history and applications of the dew point are given in the Bulletin of the American Meteorological Society.

For temperatures in degrees Fahrenheit, these approximations work out to

$$T_{dp:f} \approx T_f - \frac{9}{25}(100 - RH);$$

and

$$RH \approx 100 - \frac{25}{9}(T_f - T_{dp:f});$$

For example, a relative humidity of 100% means dew point is the same as air temp. For 90% RH, dew point is 3 degrees Fahrenheit lower than air temp. For every 10 percent lower, dew point drops 3 °F.

Frost Point

The frost point is similar to the dew point, in that it is the temperature to which a given parcel of humid air must be cooled, at constant barometric pressure, for water vapor to be deposited on a surface as ice without going through the liquid phase. (Compare with sublimation.) The frost point for a given parcel of air is always higher than the dew point, as the stronger bonding between water molecules on the surface of ice requires higher temperature to break.

Precipitation

In meteorology, precipitation is any product of the condensation of atmospheric water vapor that falls under gravity. The main forms of precipitation include drizzle, rain, sleet, snow, graupel and hail. Precipitation occurs when a portion of the atmosphere becomes saturated with water vapor, so that the water condenses and "precipitates". Thus, fog and mist are not precipitation but suspensions, because the water vapor does not condense sufficiently to precipitate. Two processes, possibly acting together, can lead to air becoming saturated: cooling the air or adding water vapor to the air. Precipitation forms as smaller droplets coalesce via collision with other rain drops or ice crystals within a cloud. Short, intense periods of rain in scattered locations are called "showers."

Moisture overriding associated with weather fronts is an overall major method of precipitation production. If enough moisture and upward motion is present, precipitation falls from convective clouds such as cumulonimbus and can organize into narrow rainbands. Where relatively warm water bodies are present, for example due to water evaporation from lakes, lake-effect snowfall becomes a concern downwind of the warm lakes within the cold cyclonic flow around the backside of extratropical cyclones. Lake-effect snowfall can be locally heavy. Thundersnow is possible within a cyclone's comma head and within lake effect precipitation bands. In mountainous areas, heavy precipitation is possible where upslope flow is maximized within windward sides of the terrain at elevation. On the leeward side of mountains, desert climates can exist due to the dry air caused by compressional heating. The movement of the monsoon trough, or intertropical convergence zone, brings rainy seasons to savannah climes.

Precipitation is a major component of the water cycle, and is responsible for depositing the fresh water on the planet. Approximately 505,000 cubic kilometres (121,000 cu mi) of water falls as precipitation each year; 398,000 cubic kilometres (95,000 cu mi) of it over the oceans and 107,000 cubic kilometres (26,000 cu mi) over land. Given the Earth's surface area, that means the globally averaged annual precipitation is 990 millimetres (39 in), but over land it is only 715 millimetres (28.1 in). Climate classification systems such as the Köppen climate classification system use average annual rainfall to help differentiate between differing climate regimes.

Precipitation may occur on other celestial bodies, e.g. when it gets cold, Mars has precipitation which most likely takes the form of ice needles, rather than rain or snow.

Types

Precipitation is a major component of the water cycle, and is responsible for depositing most of the fresh water on the planet. Approximately 505,000 km (121,000 mi) of water falls as precipitation each year, 398,000 km (95,000 cu mi) of it over the oceans. Given the Earth's surface area, that means the globally averaged annual precipitation is 990 millimetres (39 in).

Mechanisms of producing precipitation include convective, stratiform, and orographic rainfall. Convective processes involve strong vertical motions that can cause the overturning of the atmosphere in that location within an hour and cause heavy precipitation, while stratiform processes involve weaker upward motions and less intense precipitation. Precipitation can be divided into three categories, based on whether it falls as liquid water, liquid water that freezes on contact with the surface, or ice. Mixtures of different types of precipitation, including types in different categories, can fall simultaneously. Liquid forms of precipitation include rain and drizzle. Rain or drizzle that freezes on contact within a subfreezing air mass is called "freezing rain" or "freezing drizzle". Frozen forms of precipitation include snow, ice needles, ice pellets, hail, and graupel.

A thunderstorm with heavy precipitation

How the Air Becomes Saturated

Cooling Air to its Dew Point

The dew point is the temperature to which a parcel must be cooled in order to become saturated, and (unless super-saturation occurs) condenses to water. Water vapour normally begins to condense on condensation nuclei such as dust, ice, and salt in order to form clouds. An elevated portion of a frontal zone forces broad areas of lift, which form clouds decks such as altostratus or cirrostratus. Stratus is a stable cloud deck which tends to form when a cool, stable air mass is trapped underneath a warm air mass. It can also form due to the lifting of advection fog during breezy conditions.

Late-summer rainstorm in Denmark

There are four main mechanisms for cooling the air to its dew point: adiabatic cooling, conductive cooling, radiational cooling, and evaporative cooling. Adiabatic cooling occurs when air rises and expands. The air can rise due to convection, large-scale atmospheric motions, or a physical barrier such as a mountain (orographic lift). Conductive cooling occurs when the air comes into contact with a colder surface, usually by being blown from one surface to another, for example from a liquid water surface to colder land. Radiational cooling occurs due to the emission of infrared radiation, either by the air or by the surface underneath. Evaporative cooling occurs when moisture is added to the air through evaporation, which forces the air temperature to cool to its wet-bulb temperature, or until it reaches saturation.

Adding Moisture to The Air

The main ways water vapour is added to the air are: wind convergence into areas of upward motion, precipitation or virga falling from above, daytime heating evaporating water from the surface of oceans, water bodies or wet land, transpiration from plants, cool or dry air moving over warmer water, and lifting air over mountains.

Formation

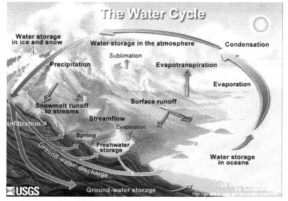
Condensation and coalescence are important parts of the water cycle.

Raindrops

Coalescence occurs when water droplets fuse to create larger water droplets, or when water droplets freeze onto an ice crystal, which is known as the Bergeron process. The fall rate of very small droplets is negligible, hence clouds do not fall out of the sky; precipitation will only occur when these coalesce into larger drops. When air turbulence occurs, water droplets collide, producing larger droplets. As these larger water droplets descend, coalescence continues, so that drops become heavy enough to overcome air resistance and fall as rain.

Raindrops have sizes ranging from 0.1 millimetres (0.0039 in) to 9 millimetres (0.35 in) mean diameter, above which they tend to break up. Smaller drops are called cloud droplets, and their shape is spherical. As a raindrop increases in size, its shape becomes more oblate, with its largest cross-section facing the oncoming airflow. Contrary to the cartoon pictures of raindrops, their shape does not resemble a teardrop. Intensity and duration of rainfall are usually inversely related, i.e., high intensity storms are likely to be of short duration and low intensity storms can have a long duration. Rain drops associated with melting hail tend to be larger than other rain drops. The METAR code for rain is RA, while the coding for rain showers is SHRA.

Ice Pellets

An accumulation of ice pellets

Ice pellets or sleet are a form of precipitation consisting of small, translucent balls of ice. Ice pellets are usually (but not always) smaller than hailstones. They often bounce when they hit the ground, and generally do not freeze into a solid mass unless mixed with freezing rain. The METAR code for ice pellets is PL.

Ice pellets form when a layer of above-freezing air exists with sub-freezing air both above and below. This causes the partial or complete melting of any snowflakes falling through the warm layer. As they fall back into the sub-freezing layer closer to the surface, they re-freeze into ice pellets. However, if the sub-freezing layer beneath the warm layer is too small, the precipitation will not have time to re-freeze, and freezing rain will be the result at the surface. A temperature profile showing a warm layer above the ground is most likely to be found in advance of a warm front during the cold season, but can occasionally be found behind a passing cold front.

Hail

A large hailstone, about 6 centimetres (2.4 in) in diameter

Like other precipitation, hail forms in storm clouds when supercooled water droplets freeze on contact with condensation nuclei, such as dust or dirt. The storm's updraft blows the hailstones to the upper part of the cloud. The updraft dissipates and the hailstones fall down, back into the updraft, and are lifted again. Hail has a diameter of 5 millimetres (0.20 in) or more. Within METAR code, GR is used to indicate larger hail, of a diameter of at least 6.4 millimetres (0.25 in). GR is derived from the French word grêle. Smaller-sized hail, as well as snow pellets, use the coding of GS, which is short for the French word grésil. Stones just larger than golf ball-sized are one of the most frequently reported hail sizes. Hailstones can grow to 15 centimetres (6 in) and weigh more than 500 grams (1 lb). In large hailstones, latent heat released by further freezing may melt the outer shell of the hailstone. The hailstone then may undergo 'wet growth', where the liquid outer shell collects other smaller hailstones. The hailstone gains an ice layer and grows increasingly larger with each ascent. Once a hailstone becomes too heavy to be supported by the storm's updraft, it falls from the cloud.

Snowflakes

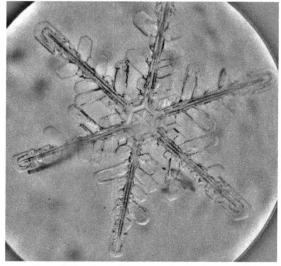

Snowflake viewed in an optical microscope

Snow crystals form when tiny supercooled cloud droplets (about 10 μm in diameter) freeze. Once a droplet has frozen, it grows in the supersaturated environment. Because water droplets are more numerous than the ice crystals the crystals are able to grow to hundreds of micrometers in size at the expense of the water droplets. This process is known as the Wegener–Bergeron–Findeisen process. The corresponding depletion of water vapour causes the droplets to evaporate, meaning that the ice crystals grow at the droplets' expense. These large crystals are an efficient source of precipitation, since they fall through the atmosphere due to their mass, and may collide and stick together in clusters, or aggregates. These aggregates are snowflakes, and are usually the type of ice particle that falls to the ground. Guinness World Records list the world's largest snowflakes as those of January 1887 at Fort Keogh, Montana; allegedly one measured 38 cm (15 inches) wide. The exact details of the sticking mechanism remain a subject of research.

Although the ice is clear, scattering of light by the crystal facets and hollows/imperfections mean that the crystals often appear white in color due to diffuse reflection of the whole spectrum of light by the small ice particles. The shape of the snowflake is determined broadly by the temperature and humidity at which it is formed. Rarely, at a temperature of around −2 °C (28 °F), snowflakes can form in threefold symmetry—triangular snowflakes. The most common snow particles are visibly irregular, although near-perfect snowflakes may be more common in pictures because they are more visually appealing. No two snowflakes are alike, which grow at different rates and in different patterns depending on the changing temperature and humidity within the atmosphere that the snowflake falls through on its way to the ground. The METAR code for snow is SN, while snow showers are coded SHSN.

Diamond Dust

Diamond dust, also known as ice needles or ice crystals, forms at temperatures approaching −40 °C (−40 °F) due to air with slightly higher moisture from aloft mixing with colder, surface based air. They are made of simple ice crystals that are hexagonal in shape. The METAR identifier for diamond dust within international hourly weather reports is IC.

Causes

Frontal Activity

Stratiform or dynamic precipitation occurs as a consequence of slow ascent of air in synoptic systems (on the order of cm/s), such as over surface cold fronts, and over and ahead of warm fronts. Similar ascent is seen around tropical cyclones outside of the eyewall, and in comma-head precipitation patterns around mid-latitude cyclones. A wide variety of weather can be found along an occluded front, with thunderstorms possible, but usually their passage is associated with a drying of the air mass. Occluded fronts usually form around mature low-pressure areas. Precipitation may occur on celestial bodies other than Earth. When it gets cold, Mars has precipitation that most likely takes the form of ice needles, rather than rain or snow.

Convection

Convective rain, or showery precipitation, occurs from convective clouds, e.g., cumulonimbus or cumulus congestus. It falls as showers with rapidly changing intensity. Convective precipitation falls over a certain area for a relatively short time, as convective clouds have limited horizontal extent. Most precipitation in the tropics appears to be convective; however, it has been suggested that stratiform precipitation also occurs. Graupel and hail indicate convection. In mid-latitudes, convective precipitation is intermittent and often associated with baroclinic boundaries such as cold fronts, squall lines, and warm fronts.

Convective precipitation

Orographic Effects

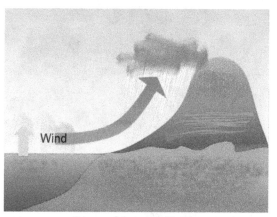

Orographic precipitation

Orographic precipitation occurs on the windward side of mountains and is caused by the rising air motion of a large-scale flow of moist air across the mountain ridge, resulting in adiabatic cooling and condensation. In mountainous parts of the world subjected to relatively consistent winds (for example, the trade winds), a more moist climate usually prevails on the windward side of a mountain than on the leeward or downwind side. Moisture is removed by orographic lift, leaving drier air on the descending and generally warming, leeward side where a rain shadow is observed.

In Hawaii, Mount Wai'ale'ale, on the island of Kauai, is notable for its extreme rainfall, as it has the second highest average annual rainfall on Earth, with 12,000 millimetres (460 in). Storm systems affect the state with heavy rains between October and March. Local climates vary considerably on each island due to their topography, divisible into windward (*Ko'olau*) and leeward (*Kona*) regions based upon location relative to the higher mountains. Windward sides face the east to northeast trade winds and receive much more rainfall; leeward sides are drier and sunnier, with less rain and less cloud cover.

In South America, the Andes mountain range blocks Pacific moisture that arrives in that continent, resulting in a desertlike climate just downwind across western Argentina. The Sierra Nevada range creates the same effect in North America forming the Great Basin and Mojave Deserts.

Snow

Extratropical cyclones can bring cold and dangerous conditions with heavy rain and snow with winds exceeding 119 km/h (74 mph), (sometimes referred to as windstorms in Europe). The band of precipitation that is associated with their warm front is often extensive, forced by weak upward vertical motion of air over the frontal boundary which condenses as it cools and produces precipitation within an elongated band, which is wide and stratiform, meaning falling out of nimbostratus clouds. When moist air tries to dislodge an arctic air mass, overrunning snow can result within the poleward side of the elongated precipitation band. In the Northern Hemisphere, poleward is towards the North Pole, or north. Within the Southern Hemisphere, poleward is towards the South Pole, or south.

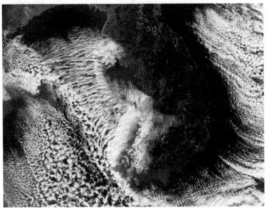
Lake-effect snow bands near the Korean Peninsula

Southwest of extratropical cyclones, curved cyclonic flow bringing cold air across the relatively warm water bodies can lead to narrow lake-effect snow bands. Those bands bring strong localized snowfall which can be understood as follows: Large water bodies such as lakes efficiently store heat that results in significant temperature differences (larger than 13 °C or 23 °F) between the water surface and the air above. Because of this temperature difference, warmth and moisture are transported upward, condensing into vertically oriented clouds which produce snow showers. The temperature decrease with height and cloud depth are directly affected by both the water temperature and the large-scale environment. The stronger the temperature decrease with height, the deeper the clouds get, and the greater the precipitation rate becomes.

In mountainous areas, heavy snowfall accumulates when air is forced to ascend the mountains and squeeze out precipitation along their windward slopes, which in cold conditions, falls in the form of snow. Because of the ruggedness of terrain, forecasting the location of heavy snowfall remains a significant challenge.

Within the Tropics

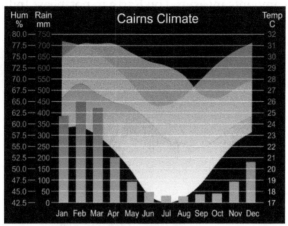

Rainfall distribution by month in Cairns showing the extent of the wet season at that location

The wet, or rainy, season is the time of year, covering one or more months, when most of the average annual rainfall in a region falls. The term *green season* is also sometimes used as a euphemism by tourist authorities. Areas with wet seasons are dispersed across portions of the tropics and subtropics. Savanna climates and areas with monsoon regimes have wet summers and dry winters. Tropical rainforests technically do not have dry or wet seasons, since their rainfall is equally distributed through the year. Some areas with pronounced rainy seasons will see a break in rainfall mid-season when the intertropical convergence zone or monsoon trough move poleward of their location during the middle of the warm season. When the wet season occurs during the warm season, or summer, rain falls mainly during the late afternoon and early evening hours. The wet season is a time when air quality improves, freshwater quality improves, and vegetation grows significantly. Soil nutrients diminish and erosion increases. Animals have adaptation and survival strategies for the wetter regime. The previous dry season leads to food shortages into the wet season, as the crops have yet to mature. Developing countries have noted that their populations show seasonal weight fluctuations due to food shortages seen before the first harvest, which occurs late in the wet season.

Tropical cyclones, a source of very heavy rainfall, consist of large air masses several hundred miles across with low pressure at the centre and with winds blowing inward towards the centre in either a clockwise direction (southern hemisphere) or counterclockwise (northern hemisphere). Although cyclones can take an enormous toll in lives and personal property, they may be important factors in the precipitation regimes of places they impact, as they may bring much-needed precipitation to otherwise dry regions. Areas in their path can receive a year's worth of rainfall from a tropical cyclone passage.

Large-scale Geographical Distribution

On the large scale, the highest precipitation amounts outside topography fall in the tropics, closely tied to the Intertropical Convergence Zone, itself the ascending branch of the Hadley cell. Mountainous locales near the equator in Colombia are amongst the wettest places on Earth. North and south of this are regions of descending air that form subtropical ridges where precipitation is low; the land surface underneath is usually arid, which forms most of the Earth's deserts. An exception to this rule is in Hawaii, where upslope flow due to the trade winds lead to one of the wettest locations on Earth. Otherwise, the flow of the Westerlies into the Rocky Mountains lead to the wettest, and at elevation snowiest, locations within North America. In Asia during the wet season, the flow of moist air into the Himalayas leads to some of the greatest rainfall amounts measured on Earth in northeast India.

Measurement

Standard rain gauge

The standard way of measuring rainfall or snowfall is the standard rain gauge, which can be found in 100 mm (4 in) plastic and 200 mm (8 in) metal varieties. The inner cylinder is filled by 25 mm (1 in) of rain, with overflow flowing into the outer cylinder. Plastic gauges have markings on the inner cylinder down to 0.25 mm (0.01 in) resolution, while metal gauges require use of a stick designed with the appropriate 0.25 mm (0.01 in) markings. After the inner cylinder is filled, the amount inside it is discarded, then filled with the remaining rainfall in the outer cylinder until all the fluid in the outer cylinder is gone, adding to the overall total until the outer cylinder is empty. These gauges are used in the winter by removing the funnel and inner cylinder and allowing snow and freezing rain to collect inside the outer cylinder. Some add anti-freeze to their gauge so they do not have to melt the snow or ice that falls into the gauge. Once the snowfall/ice is finished accumulating, or as 300 mm (12 in) is approached, one can either bring it inside to melt, or use lukewarm water to fill the inner cylinder with in order to melt the frozen precipitation in the outer cylinder, keeping track of the warm fluid added, which is subsequently subtracted from the overall total once all the ice/snow is melted.

Other types of gauges include the popular wedge gauge (the cheapest rain gauge and most fragile), the tipping bucket rain gauge, and the weighing rain gauge. The wedge and tipping bucket gauges will have problems with snow. Attempts to compensate for snow/ice by warming the tipping bucket meet with limited success, since snow may sublimate if the gauge is kept much above freezing. Weighing gauges with antifreeze should do fine with snow, but again, the funnel needs to be removed before the event begins. For those looking to measure rainfall the most inexpensively, a can that is cylindrical with straight sides will act as a rain gauge if left out in the open, but its accuracy will depend on what ruler is used to measure the rain with. Any of the above rain gauges can be made at home, with enough know-how.

When a precipitation measurement is made, various networks exist across the United States and elsewhere where rainfall measurements can be submitted through the Internet, such as CoCo-

RAHS or GLOBE. If a network is not available in the area where one lives, the nearest local weather office will likely be interested in the measurement.

Hydrometeor Defi nition

A concept used in precipitation measurement is the hydrometeor. Bits of liquid or solid water in the atmosphere are known as hydrometeors. Formations due to condensation, such as clouds, haze, fog, and mist, are composed of hydrometeors. All precipitation types are made up of hydrometeors by definition, including virga, which is precipitation which evaporates before reaching the ground. Particles blown from the Earth's surface by wind, such as blowing snow and blowing sea spray, are also hydrometeors.

Satellite Estimates

Although surface precipitation gauges are considered the standard for measuring precipitation, there are many areas in which their use is not feasible. This includes the vast expanses of ocean and remote land areas. In other cases, social, technical or administrative issues prevent the dissemination of gauge observations. As a result, the modern global record of precipitation largely depends on satellite observations.

Satellite sensors work by remotely sensing precipitation—recording various parts of the electromagnetic spectrum that theory and practice show are related to the occurrence and intensity of precipitation. The sensors are almost exclusively passive, recording what they see, similar to a camera, in contrast to active sensors (radar, lidar) that send out a signal and detect its impact on the area being observed.

Satellite sensors now in practical use for precipitation fall into two categories. Thermal infrared (IR) sensors record a channel around 11 micron wavelength and primarily give information about cloud tops. Due to the typical structure of the atmosphere, cloud-top temperatures are approximately inversely related to cloud-top heights, meaning colder clouds almost always occur at higher altitudes. Further, cloud tops with a lot of small-scale variation are likely to be more vigorous than smooth-topped clouds. Various mathematical schemes, or algorithms, use these and other properties to estimate precipitation from the IR data.

The second category of sensor channels is in the microwave part of the electromagnetic spectrum. The frequencies in use range from about 10 gigahertz to a few hundred GHz. Channels up to about 37 GHz primarily provide information on the liquid hydrometeors (rain and drizzle) in the lower parts of clouds, with larger amounts of liquid emitting higher amounts of microwave radiant energy. Channels above 37 GHz display emission signals, but are dominated by the action of solid hydrometeors (snow, graupel, etc.) to scatter microwave radiant energy. Satellites such as the Tropical Rainfall Measuring Mission (TRMM) and the Global Precipitation Measurement (GPM) mission employ microwave sensors to form precipitation estimates.

Additional sensor channels and products have been demonstrated to provide additional useful information including visible channels, additional IR channels, water vapor channels and atmospheric sounding retrievals. However, most precipitation data sets in current use do not employ these data sources.

Satellite Data Sets

The IR estimates have rather low skill at short time and space scales, but are available very frequently (15 minutes or more often) from satellites in geosynchronous Earth orbit. IR works best in cases of deep, vigorous convection—such as the tropics—and becomes progressively less useful in areas where stratiform (layered) precipitation dominates, especially in mid- and high-latitude regions. The more-direct physical connection between hydrometeors and microwave channels gives the microwave estimates greater skill on short time and space scales than is true for IR. However, microwave sensors fly only on low Earth orbit satellites, and there are few enough of them that the average time between observations exceeds three hours. This several-hour interval is insufficient to adequately document precipitation because of the transient nature of most precipitation systems as well as the inability of a single satellite to appropriately capture the typical daily cycle of precipitation at a given location.

Since the late 1990s, several algorithms have been developed to combine precipitation data from multiple satellites' sensors, seeking to emphasize the strengths and minimize the weaknesses of the individual input data sets. The goal is to provide "best" estimates of precipitation on a uniform time/space grid, usually for as much of the globe as possible. In some cases the long-term homogeneity of the dataset is emphasized, which is the Climate Data Record standard.

In other cases, the goal is producing the best instantaneous satellite estimate, which is the High Resolution Precipitation Product approach. In either case, of course, the less-emphasized goal is also considered desirable. One key result of the multi-satellite studies is that including even a small amount of surface gauge data is very useful for controlling the biases that are endemic to satellite estimates. The difficulties in using gauge data are that 1) their availability is limited, as noted above, and 2) the best analyses of gauge data take two months or more after the observation time to undergo the necessary transmission, assembly, processing and quality control. Thus, precipitation estimates that include gauge data tend to be produced further after the observation time than the no-gauge estimates. As a result, while estimates that include gauge data may provide a more accurate depiction of the "true" precipitation, they are generally not suited for real- or near-real-time applications.

The work described has resulted in a variety of datasets possessing different formats, time/space grids, periods of record and regions of coverage, input datasets, and analysis procedures, as well as many different forms of dataset version designators. In many cases, one of the modern multi-satellite data sets is the best choice for general use.

Return Period

The likelihood or probability of an event with a specified intensity and duration, is called the return period or frequency. The intensity of a storm can be predicted for any return period and storm duration, from charts based on historic data for the location. The term *1 in 10 year storm* describes a rainfall event which is rare and is only likely to occur once every 10 years, so it has a 10 percent likelihood any given year. The rainfall will be greater and the flooding will be worse than the worst storm expected in any single year. The term *1 in 100 year storm* describes a rainfall event which is extremely rare and which will occur with a likelihood of only once in a century, so has a 1 percent likelihood in any given year. The rainfall will be extreme and flooding to be worse than a 1 in 10 year event. As with all probability events, it is possible though unlikely to have two "1 in 100 Year Storms" in a single year.

Role in Köppen Climate Classification

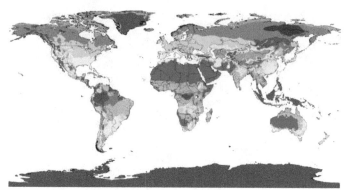

Updated Köppen-Geiger climate map

The Köppen classification depends on average monthly values of temperature and precipitation. The most commonly used form of the Köppen classification has five primary types labeled A through E. Specifically, the primary types are A, tropical; B, dry; C, mild mid-latitude; D, cold mid-latitude; and E, polar. The five primary classifications can be further divided into secondary classifications such as rain forest, monsoon, tropical savanna, humid subtropical, humid continental, oceanic climate, Mediterranean climate, steppe, subarctic climate, tundra, polar ice cap, and desert.

Rain forests are characterized by high rainfall, with definitions setting minimum normal annual rainfall between 1,750 and 2,000 mm (69 and 79 in). A tropical savanna is a grassland biome located in semi-arid to semi-humid climate regions of subtropical and tropical latitudes, with rainfall between 750 and 1,270 mm (30 and 50 in) a year. They are widespread on Africa, and are also found in India, the northern parts of South America, Malaysia, and Australia. The humid subtropical climate zone is where winter rainfall (and sometimes snowfall) is associated with large storms that the westerlies steer from west to east. Most summer rainfall occurs during thunderstorms and from occasional tropical cyclones. Humid subtropical climates lie on the east side continents, roughly between latitudes 20° and 40° degrees away from the equator.

An oceanic (or maritime) climate is typically found along the west coasts at the middle latitudes of all the world's continents, bordering cool oceans, as well as southeastern Australia, and is accompanied by plentiful precipitation year round. The Mediterranean climate regime resembles the climate of the lands in the Mediterranean Basin, parts of western North America, parts of Western and South Australia, in southwestern South Africa and in parts of central Chile. The climate is characterized by hot, dry summers and cool, wet winters. A steppe is a dry grassland. Subarctic climates are cold with continuous permafrost and little precipitation.

Effect on Agriculture

Precipitation, especially rain, has a dramatic effect on agriculture. All plants need at least some water to survive, therefore rain (being the most effective means of watering) is important to agriculture. While a regular rain pattern is usually vital to healthy plants, too much or too little rainfall can be harmful, even devastating to crops. Drought can kill crops and increase erosion, while overly wet weather can cause harmful fungus growth. Plants need varying amounts of rainfall to survive. For example, certain cacti require small amounts of water, while tropical plants may need up to hundreds of inches of rain per year to survive.

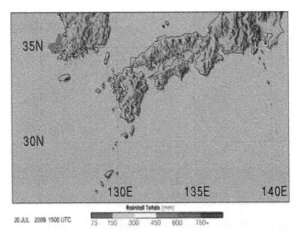

Rainfall estimates for southern Japan and the surrounding region from July 20 to 27, 2009.

In areas with wet and dry seasons, soil nutrients diminish and erosion increases during the wet season. Animals have adaptation and survival strategies for the wetter regime. The previous dry season leads to food shortages into the wet season, as the crops have yet to mature. Developing countries have noted that their populations show seasonal weight fluctuations due to food shortages seen before the first harvest, which occurs late in the wet season.

Changes Due to Global Warming

Increasing temperatures tend to increase evaporation which leads to more precipitation. Precipitation has generally increased over land north of 30°N from 1900 to 2005 but has declined over the tropics since the 1970s. Globally there has been no statistically significant overall trend in precipitation over the past century, although trends have varied widely by region and over time. Eastern portions of North and South America, northern Europe, and northern and central Asia have become wetter. The Sahel, the Mediterranean, southern Africa and parts of southern Asia have become drier. There has been an increase in the number of heavy precipitation events over many areas during the past century, as well as an increase since the 1970s in the prevalence of droughts—especially in the tropics and subtropics. Changes in precipitation and evaporation over the oceans are suggested by the decreased salinity of mid- and high-latitude waters (implying more precipitation), along with increased salinity in lower latitudes (implying less precipitation, more evaporation, or both). Over the contiguous United States, total annual precipitation increased at an average rate of 6.1% per century since 1900, with the greatest increases within the East North Central climate region (11.6% per century) and the South (11.1%). Hawaii was the only region to show a decrease (-9.25%).

Changes Due to Urban Heat Island

The urban heat island warms cities 0.6 to 5.6 °C (1.1 to 10.1 °F) above surrounding suburbs and rural areas. This extra heat leads to greater upward motion, which can induce additional shower and thunderstorm activity. Rainfall rates downwind of cities are increased between 48% and 116%. Partly as a result of this warming, monthly rainfall is about 28% greater between 32 to 64 kilometres (20 to 40 mi) downwind of cities, compared with upwind. Some cities induce a total precipitation increase of 51%.

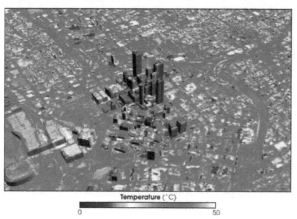

Image of Atlanta, Georgia, showing temperature distribution, with hot areas appearing white

Forecasting

The Quantitative Precipitation Forecast (abbreviated QPF) is the expected amount of liquid precipitation accumulated over a specified time period over a specified area. A QPF will be specified when a measurable precipitation type reaching a minimum threshold is forecast for any hour during a QPF valid period. Precipitation forecasts tend to be bound by synoptic hours such as 0000, 0600, 1200 and 1800 GMT. Terrain is considered in QPFs by use of topography or based upon climatological precipitation patterns from observations with fine detail. Starting in the mid to late 1990s, QPFs were used within hydrologic forecast models to simulate impact to rivers throughout the United States. Forecast models show significant sensitivity to humidity levels within the planetary boundary layer, or in the lowest levels of the atmosphere, which decreases with height. QPF can be generated on a quantitative, forecasting amounts, or a qualitative, forecasting the probability of a specific amount, basis. Radar imagery forecasting techniques show higher skill than model forecasts within six to seven hours of the time of the radar image. The forecasts can be verified through use of rain gauge measurements, weather radar estimates, or a combination of both. Various skill scores can be determined to measure the value of the rainfall forecast.

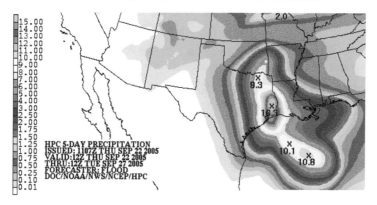

Example of a five-day rainfall forecast from the Hydrometeorological Prediction Center

References

- Fredlund, D.G.; Rahardjo, H. (1993). Soil Mechanics for Unsaturated Soils (PDF). Wiley-Interscience. ISBN 978-0-471-85008-3. OCLC 26543184. Retrieved 2008-05-21.

- Mark Zachary Jacobson (2005). Fundamentals of Atmospheric Modeling (2nd ed.). Cambridge University Press. ISBN 0-521-83970-X. OCLC 243560910.

- "Famine in Scotland: The 'Ill Years' of the 1690s". Karen Cullen,Karen J. Cullen (2010). Edinburgh University Press. p.21. ISBN 0-7486-3887-3

- C. Donald Ahrens (2007). Meteorology today: an introduction to weather, climate, and the environment. Cengage Learning. p. 296. ISBN 978-0-495-01162-0. Retrieved 2010-12-17.

- International Civil Aviation Organization. Manual of the ICAO Standard Atmosphere, Doc 7488-CD, Third Edition, 1993. ISBN 92-9194-004-6.

- John M. Wallace; Peter V. Hobbs (24 March 2006). Atmospheric Science: An Introductory Survey. Academic Press. pp. 83–. ISBN 978-0-08-049953-6.

- H. Edward Reiley; Carroll L. Shry (2002). Introductory horticulture. Cengage Learning. p. 40. ISBN 978-0-7668-1567-4. Retrieved 2011-01-31.

- Jennifer E. Lawson (2001). Hands-on Science: Light, Physical Science (matter) - Chapter 5: The Colors of Light. Portage & Main Press. p. 39. ISBN 978-1-894110-63-1. Retrieved 2009-06-28.

- Michael A. Mares (1999). Encyclopedia of Deserts. University of Oklahoma Press. p. 252. ISBN 978-0-8061-3146-7. Retrieved 2009-01-02.

- C. D. Haynes; M. G. Ridpath; M. A. J. Williams (1991). Monsoonal Australia. Taylor & Francis. p. 90. ISBN 978-90-6191-638-3. Retrieved 2008-12-27.

- A. Roberto Frisancho (1993). Human Adaptation and Accommodation. University of Michigan Press, pp. 388. ISBN 978-0-472-09511-7. Retrieved on 2008-12-27

- Planton, Serge (France; editor) (2013). "Annex III. Glossary: IPCC - Intergovernmental Panel on Climate Change" (PDF). IPCC Fifth Assessment Report. p. 1450. Retrieved 25 July 2016.

- Brown, Dwayne; Cabbage, Michael; McCarthy, Leslie; Norton, Karen (20 January 2016). "NASA, NOAA Analyses Reveal Record-Shattering Global Warm Temperatures in 2015". NASA. Retrieved 21 January 2016.

- Shepherd, Dr. J. Marshall; Shindell, Drew; O'Carroll, Cynthia M. (1 February 2005). "What's the Difference Between Weather and Climate?". NASA. Retrieved 13 November 2015.

- "Commission For Climatology: Over Eighty Years of Service" (pdf). World Meteorological Organization. 2011. pp. 6, 8, 10, 21, 26. Retrieved 1 September 2015.

- Chris Landsea (2010-04-21). "Subject: E1), Which is the most intense tropical cyclone on record?". Atlantic Oceanographic and Meteorological Laboratory. Archived from the original on 6 December 2010. Retrieved 2010-11-23.

- Kenneth G. Libbrecht (2001). "Morphogenesis on Ice: The Physics of Snow Crystals" (PDF). Engineering & Science. California Institute of Technology (1): 12. Retrieved 2010-01-21.

- Goddard Space Flight Center (2002-06-18). "NASA Satellite Confirms Urban Heat Islands Increase Rainfall Around Cities". National Aeronautics and Space Administration. Archived from the original on March 16, 2010. Retrieved 2009-07-17.

Understanding Important Meteorological Phenomenon

Most atmospheric phenomenon are a result of changes in temperature, moisture boundary and moisture instability. Cloud formation, rain, storm and cyclones are some of the phenomenon explored in this chapter. There is a section dedicated to cyclogenesis to help the reader form a deeper understanding of the formation and/or strengthening of cyclonic circulation.

Cloud

Cumuliform cloudscape over Swifts Creek, Australia

In meteorology, a cloud is an aerosol comprising a visible mass of minute liquid droplets or frozen crystals, both of which are made of water or various chemicals. The droplets or particles are suspended in the atmosphere above the surface of a planetary body. On Earth, clouds are formed by the saturation of air in the homosphere (which includes the troposphere, stratosphere, and mesosphere). The air may be cooled to its dew point by a variety of atmospheric processes or it may gain moisture (usually in the form of water vapor) from an adjacent source. Nephology is the science of clouds which is undertaken in the cloud physics branch of meteorology.

Cloud types in the troposphere, the atmospheric layer closest to Earth's surface, have Latin names due to the universal adaptation of Luke Howard's nomenclature. It was formally proposed in December 1802 and published for the first time the following year. It became the basis of a modern international system that classifies these tropospheric aerosols into five physical *forms* and three altitude levels or *étages*. These physical types, in approximate descending order of mean altitude range, include *cirriform* wisps and patches, *stratocumuliform* layers (mainly structured as rolls, ripples, and patches), *cumuliform* heaps and tufts, *stratiform* sheets, and very large *cumulonimbiform* heaps that often show complex structure. The physical forms are cross-classified by altitude levels to produce ten basic genus-types or *genera*. Some of these basic types are common to more than one form or more than one level, as illustrated in the stratocumuliform and cumuliform columns of the classification table below. Most genera can be divided into *species*, some of which are common to more than one genus. These can be subdivided into *varieties*, some of which are common to more than one genus or species.

Cirriform clouds that form higher up in the stratosphere and mesosphere have common names for their main types, but are sub-classified *alpha-numerically* rather than with the elaborate system of Latin names given to cloud types in the troposphere. They are relatively uncommon and are mostly seen in the polar regions of Earth. Clouds have been observed in the atmospheres of other planets and moons in the Solar System and beyond. However, due to their different temperature characteristics, they are often composed of other substances such as methane, ammonia, and sulfuric acid as well as water.

Classification of major types	Cirriform	Stratocumuliform	Cumuliform	Stratiform	Cumulonimbiform
Extreme level (mesosphere)	Noctilucent				
Very high level (stratosphere)	Nacreous/ non-nacreous				
High-level	Cirrus	Cirrocumulus (layered)	Cirrocumulus (tufted)	Cirrostratus	
Mid-level		Altocumulus (layered)	Altocumulus (tufted)	Altostratus	
Low-level		Stratocumulus	Cumulus (small)	Stratus	
Multi-level			Cumulus (large)	Nimbostratus	Cumulonimbus

Etymology

The origin of the term *cloud* can be found in the old English *clud* or *clod*, meaning a hill or a mass of rock. Around the beginning of the 13th century, it was extended as a metaphor to include rain clouds as masses of evaporated water in the sky because of the similarity in appearance between a mass of rock and a cumulus heap cloud. Over time, the metaphoric term replaced the original old English *weolcan* to refer to clouds in general.

History of Cloud Science and Nomenclature

Aristotle and Theophrastus

Ancient cloud studies were not made in isolation, but were observed in combination with other weather elements and even other natural sciences. In about 340 BC the Greek philosopher Aristotle wrote *Meteorologica*, a work which represented the sum of knowledge of the time about natural science, including weather and climate. For the first time, precipitation and the clouds from which precipitation fell were called meteors. From that word came the modern term meteorology, the study of clouds and weather. Meteorologica was based on intuition and simple observation, but not on what is now considered the scientific method. Nevertheless, it was the first known work that attempted to treat a broad range of meteorological topics.

The magazine *De Mundo* (attributed to Pseudo-Aristotle) noted:

Cloud is a vaporous mass, concentrated and producing water. Rain is produced from the compression of a closely condensed cloud, varying according to the pressure exerted on the cloud; when the pressure is slight it scatters gentle drops; when it is great it produces a more violent fall, and we call this a shower, being heavier than ordinary rain, and forming continuous masses of water falling over earth. Snow is produced by the breaking up of condensed clouds, the cleavage taking place before the change into water; it is the process of cleavage which causes its resemblance to foam and its intense whiteness, while the cause of its coldness is the congelation of the moisture in it before it is dispersed or rarefied. When snow is violent and falls heavily we call it a blizzard. Hail is produced when snow becomes densified and acquires impetus for a swifter fall from its close mass; the weight becomes greater and the fall more violent in proportion to the size of the broken fragments of cloud. Such then are the phenomena which occur as the result of moist exhalation.

Several years after Aristotle's book, his pupil Theophrastus put together a book on weather forecasting called *The Book of Signs*. Various indicators such as solar and lunar halos formed by high clouds were presented as ways to forecast the weather. The combined works of Aristotle and Theophrastus had such authority they became the main influence in the study of clouds, weather and weather forecasting for nearly 2000 years.

Luke Howard, Jean-Baptiste Lamarck, and The First Comprehensive Classification

After centuries of speculative theories about the formation and behavior of clouds, the first truly scientific studies were undertaken by Luke Howard in England and Jean-Baptiste Lamarck in France. Howard was a methodical observer with a strong grounding in the Latin language and used his background to classify the various tropospheric cloud types during 1802. He believed that the changing cloud forms in the sky could unlock the key to weather forecasting. Lamarck had worked independently on cloud classification the same year and had come up with a different naming scheme that failed to make an impression even in his home country of France because it used unusual French names for cloud types. His system of nomenclature included twelve categories of clouds, with such names as (translated from French) hazy clouds, dappled clouds and broom-like clouds. By contrast, Howard used universally accepted Latin, which caught on quickly after it was published in 1803. As a sign of the popularity of the naming scheme, the German dramatist and poet Johann Wolfgang von Goethe composed four poems about clouds, dedicating them to Howard. An elaboration of Howard's system was eventually formally adopted by the International Meteorological Conference in 1891.

Howard's original system established three physical categories or *forms* based on appearance and process of formation: *cirriform* (mainly detached and wispy), *cumuliform* or convective (mostly detached and heaped, rolled, or rippled), and non-convective *stratiform* (mainly continuous layers in sheets). These were cross-classified into *lower* and *upper* étages. Cumuliform clouds forming in the lower level were given the genus name cumulus from the Latin word for *heap*, and low stratiform clouds the genus name stratus from the Latin word for *sheet* or *layer*. Physically similar clouds forming in the upper étage were given the genus names cirrocumulus (generally showing more limited convective activity than low level cumulus) and cirrostratus, respectively. Cirriform clouds were identified as always upper level and given the genus name cirrus from the Latin for 'fibre' or 'hair'.

Altocumulus stratiformis duplicatus at sunrise in the California Mojave Desert, USA

In addition to these individual cloud types; Howard added two names to designate cloud systems consisting of more than one form joined together or located in very close proximity. Cumulostratus described large cumulus clouds blended with stratiform layers in the lower or upper levels. The term nimbus was given to complex systems of cirriform, cumuliform, and stratiform clouds with sufficient vertical development to produce significant precipitation, and it came to be identified as a distinct *nimbiform* physical category.

Howard'S Successors

In 1840, German meteorologist Ludwig Kaemtz added stratocumulus to Howard's canon as a mostly detached low-étage genus of *limited* convection. It was defined as having cumuliform- and stratiform characteristics integrated into a single layer (in contrast to cumulostratus which was deemed to be composite in nature and could be structured into more than one layer). This led to the recognition of a *stratocumuliform* physical category that included rolled and rippled clouds classified separately from the more freely convective heaped cumuliform clouds.

During the mid 1850s, Emilien Renou, director of the Parc Saint-Maur and Montsouris observatories, began work on an elaboration of Howard's classifications that would lead to the introduction during the 1870s of altocumulus (physically more closely related to stratocumulus than to cumulus) and altostratus. These were respectively stratocumuliform and stratiform cloud genera of a newly defined *middle* étage above stratocumulus and stratus but below cirrocumulus and cirrostratus.

Middle clouds over Santa Clarita, CA. Altocumulus floccus producing virga near top and middle of image merging into altostratus translucidus near horizon.

In 1880, Philip Weilbach, secretary and librarian at the Art Academy in Copenhagen, and like Luke Howard, an amateur meteorologist, unsuccessfully proposed an alternative to Howard's classification. However, he also proposed and had accepted by the permanent committee of the International Meteorological Organization (IMO), a forerunner of the present-day World Meteorological Organization (WMO), the designation of a new free-convective vertical or multi-étage genus type, cumulonimbus, which would be distinct from cumulus and nimbus and identifiable by its often very complex structure (frequently including a cirriform top and what are now recognized as multiple accessory clouds), and its ability to produce thunder. With this addition, a canon of ten tropospheric cloud *genera* was established that came to be officially and universally accepted. Howard's cumulostratus was not included as a distinct type, having effectively been reclassified into its component cumuliform and stratiform genus types already included in the new canon.

In 1890, Otto Jesse revealed the discovery and identification of the first clouds known to form above the troposphere. He proposed the name *noctilucent* which is Latin for *night shining*. Because of the extremely high altitudes of these clouds in what is now known to be the mesosphere, they could become illuminated by the a sun's rays when the sky was nearly dark after sunset and before sunrise. Three years later, Henrik Mohn revealed a similar discovery of nacreous clouds in what is now considered the stratosphere.

In 1896, the first cloud atlas sanctioned by the IMO was produced by Teisserenc de Borte based on collaborations with Hugo H. Hildebrandsson. The latter had become the first researcher to use photography for the study and classification of clouds in 1879.

Alternatives to Howard's classification system were proposed throughout the 19th century. Heinrich Dove of Germany and Elias Loomis of the United States came up with other schemes in 1828 and 1841 respectively, but neither met with international success. Additional proposals were made by Andre Poey (1863), Clemment Ley (1894), and H.H. Clayton (1896), but their systems, like earlier alternative schemes, differed too much from Howard's to have any success beyond the adoption of some secondary cloud types. However, Clayton's idea to formalize the division of clouds by their physical structures into cirriform, stratiform, "flocciform" (stratocumuliform) and cumuliform (with the later addition of cumulonimbiform), eventually found favor as an aid in the analysis of satellite cloud images.

20th-century Developments

A further modification of the genus classification system came when an IMC commission for the study of clouds put forward a refined and more restricted definition of the genus nimbus which was effectively reclassified as a stratiform cloud type. It was then renamed nimbostratus and published with the new name in the 1932 edition of the *International Atlas of Clouds and of States of the Sky*. This left cumulonimbus as the only nimbiform type as indicated by its root-name.

On April 1, 1960, the first successful weather satellite, TIROS-1 (Television Infrared Observation Satellite), was launched from Cape Canaveral, Florida by the National Aeronautics and Space Administration (NASA) with the participation of The US Army Signal Research and Development Lab, RCA, the US Weather Bureau, and the US Naval Photographic Center. During its 78-day mission, it relayed thousands of pictures showing the structure of large-scale cloud regimes, and proved that satellites could provide useful surveillance of global weather conditions from space.

In 1976, the United Kingdom Department of Industry published a modification of the international cloud classification system adapted for satellite cloud observations. It was co-sponsored by NASA and showed a change in name of the nimbiform type to *cumulonimbiform*, although the earlier name and original meaning pertaining to all rain clouds can still be found in some classifications.

Tropospheric

Physical Forms

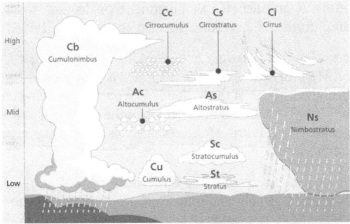

Genus classification by altitude of occurrence. Multi-etage types or sub-types not limited by altitude include cumulonimbus-Cb, cumulus congestus-Cu con (not shown), and nimbostratus-Ns.

Clouds can be divided into five physical forms based on physical structure and process of formation. These forms are commonly used for the purpose of satellite analysis.

The individual *genus* types result from the physical forms being cross-classified by altitude level or étage within the troposphere. The base-height range for each étage varies depending on the latitudinal geographical zone. A consensus exists as to the designation of high, middle, and low étages, the makeup of the basic canon of ten cloud genera that results from this cross-classification, and the étage designations of non-vertical genus types. Clouds with significant vertical extent occupy more than one étage and are commonly, but not always, treated as a separate group or sub-group, or given separate descriptions within the context of the standard étages. The physical forms and their constituent genera are given below in approximate ascending order of instability or convective activity.

Stratiform

Non-convective stratiform clouds appear in *stable* airmass conditions and, in general, have flat sheet-like structures that can form at any altitude in the troposphere. Very low stratiform cloud results when advection fog is lifted above surface level during breezy conditions. The stratiform group is cross-classified into the genera cirrostratus (high-étage), altostratus (middle-étage), stratus (low-étage), and nimbostratus (multi-étage).

Cirriform

Cirriform clouds are generally of the genus cirrus and have the appearance of detached or semi-merged filaments. They form at high tropospheric altitudes in air that is mostly stable with little or no convective activity, although denser patches may occasionally show buildups caused by *limited* high-level convection where the air is partly *unstable*.

Stratocumuliform

Clouds of this structure have both cumuliform and stratiform characteristics in the form of rolls or ripples. They generally form as a result of *limited convection* in an otherwise mostly stable airmass topped by an inversion layer. The stratocumuliform group is cross-classified into layered cirrocumulus (high-étage), layered altocumulus (middle-étage), and stratocumulus (low-étage).

Cumuliform

Cumuliform clouds generally appear in isolated heaps or tufts. They are the product of localized but generally *free-convective* lift where there are no inversion layers in the atmosphere to limit vertical growth. In general, small cumuliform clouds tend to indicate comparatively weak instability. Larger cumuliform types are a sign of moderate to strong atmospheric instability and convective activity. Depending on their vertical size, clouds of the cumulus genus-type may be low or multi-étage. Tufted altocumulus and cirrocumulus genera in the middle and high étages are also considered cumuliform because they have a more detached heaped structure than their layered stratocumuliform variants.

Cumulonimbiform

The largest free-convective clouds comprise the genus cumulonimbus which are multi-étage because of their great vertical extent. They occur in highly unstable air and often have complex structures that include cirriform tops and multiple accessory clouds.

Étages and Genera

The forms and resultant genus types can be grouped by étage. This is generally done for the purpose of cloud atlases, surface weather observations and weather maps. These maps are produced from information in the international synoptic code (or SYNOP) that is transmitted at regular intervals by professionally trained staff at major weather stations.

Non-vertical or single-étage genera are listed and summarised below in approximate descending order of the altitude at which each is normally based. Multi-étage clouds with significant vertical extent are separately listed and summarised in approximate ascending order of instability or convective activity.

High Étage

Clouds of the high étage form at altitudes of 3,000 to 7,600 m (10,000 to 25,000 ft) in the polar regions, 5,000 to 12,200 m (16,500 to 40,000 ft) in the temperate regions and 6,100 to 18,300 m (20,000 to 60,000 ft) in the tropical region. All cirriform clouds are classified as high and thus constitute a single genus *cirrus* (Ci). Stratocumuliform and stratiform clouds in the high étage carry the prefix *cirro-*, yielding the respective genus names *cirrocumulus* (Cc) and *cirrostratus* (Cs). When comparatively low-resolution satellite images of high clouds are analized without supporting data from direct human observations, it becomes impossible to distinguish between individual genus types which are then collectively identified as cirrus-type.

High cirrus uncinus and cirrus fibratus upper-left merging into cirrostratus fibratus with some higher cirrocumulus floccus upper right

- Genus cirrus (Ci):

  These are mostly fibrous wisps of delicate white cirriform ice crystal cloud that show up clearly against the blue sky. Cirrus are generally non-convective except castellanus and floccus subtypes which show limited convection. They often form along a high altitude jetstream and at the very leading edge of a frontal or low-pressure disturbance where they may merge into cirrostratus. These high clouds do not produce precipitation.

- Genus cirrocumulus (Cc):

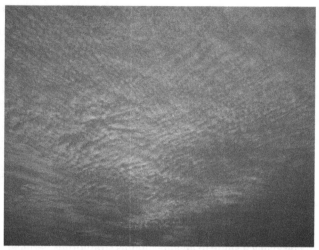

A large field of cirrocumulus stratiformis

This is most commonly a pure white high-étage stratocumuliform layer of limited convection. It is composed of ice crystals or supercooled water droplets appearing as small unshaded round masses or flakes in groups or lines with ripples like sand on a beach. Cirrocumulus occasionally forms alongside cirrus and may be accompanied or replaced by cirrostratus clouds at the very leading edge of an active weather system. Tufted cirrocumulus forms in more isolated heaps than the layered varient and can therefore be considered a cumuliform cloud which retains its pure white coloration.

- Genus cirrostratus (Cs):

Cirrostratus is a thin non-convective stratiform ice crystal veil that typically gives rise to halos caused by refraction of the sun's rays. The sun and moon are visible in clear outline. Cirrostratus often thickens into altostratus ahead of a warm front or low-pressure area.

Middle Étage

Non-vertical clouds in the middle étage are prefixed by *alto-*, yielding the genus names *altocumulus* (Ac) and *altostratus* (As). These clouds can form as low as 2,000 m (6,500 ft) above surface at any latitude, but may be based as high as 4,000 m (13,000 ft) near the poles, 7,000 m (23,000 ft) at mid latitudes, and 7,600 m (25,000 ft) in the tropics. As with high clouds, it is not always possible to distinguish between individual genera using satellite photography alone. Without the addition of human observations, these clouds are usually collectively identified as 'middle-type' on satellite images.

- Genus altocumulus (Ac):

This is most commonly a middle-étage stratocumuliform cloud layer of limited convection that is usually appears in the form of irregular patches or more extensive sheets arranges in groups, lines, or waves. High altocumulus may resemble cirrocumulus but is usually thicker and composed of water droplets so that the bases show at least some light-grey shading. Opaque altocumulus associated with a weak frontal or low-pressure disturbance can produce virga, very light intermittent precipitation that evaporates before reaching the ground. If the altocumulus is mixed with moisture-laden altostratus, the precipitation may reach the ground. As with cirrocumulus, tufted altocumulus in isolated heaps can be considered a cumuliform rather than a stratocumuliform cloud

Sunrise scene giving a shine to an altocumulus stratiformis perlucidus cloud

- Genus altostratus (As):

Altostratus translucidus near top of photo merging into altostratus opacus near bottom

Altostratus is a mid-level opaque or translucent stratiform or non-convective veil of grey/blue-grey cloud that often forms along warm fronts and around low-pressure areas. Altostratus is usually composed of water droplets but may be mixed with ice crystals at higher altitudes. Widespread opaque altostratus can produce light continuous or intermittent precipitation. Precipitation commonly becomes heavier and more widespread if it thickens into nimbostratus.

Low Étage

Low-étage clouds are found from near surface up to 2,000 m (6,500 ft). Genus types in this étage either have no prefix or carry one that refers to a characteristic other than altitude.

Stratocumulus stratiformis perlucidus over Galapagos, Tortuga Bay

- Genus stratocumulus (Sc):

  This genus type is a stratocumuliform cloud layer of limited convection, usually in the form of irregular patches or more extensive sheets similar to altocumulus but having larger elements with deeper-gray shading. Opaque stratocumulus can produce very light intermittent precipitation. This cloud often forms under a precipitating deck of altostratus or high-based nimbostratus associated with a well-developed warm front, slow-moving cold front, or low-pressure area. This can create the illusion of continuous precipitation of more than very light intensity falling from stratocumulus.

- Genus stratus (St):

At level with stratus nebulosus translucidus clouds

This is a flat or sometimes ragged non-convective stratiform type that sometimes resembles elevated fog. Only very weak precipitation can fall from this cloud (usually drizzle or snow grains), although heavier rain or snow may fall through a stratus layer from a higher precipitating cloud deck. When a low stratiform cloud contacts the ground, it is called fog if the prevailing surface visibility is less than 1 kilometer, although radiation and advection types of fog tend to form in clear air rather than from stratus layers. If the visibility increases to 1 kilometer or higher in any kind of fog, the visible condensation is termed mist.

- Genus cumulus (Cu) – *little vertical extent*:

Low cumulus humilis and some moderate vertical cumulus mediocris in the foreground and background with some stratocumulus stratiformis perlucidus clouds mainly in the foreground

These are small detached fair-weather cumuliform clouds that have nearly horizontal bases and flattened tops, and do not produce rain showers.

Multi-étage (Low to Mid-Level Cloud Base)

These clouds have low to middle-étage bases that form anywhere from near surface to about 2,400 m (8,000 ft) and tops that can extend into the high étage. The term *vertical* is often used in connection with this group and is useful for distinguishing between clouds of moderate, deep, and towering vertical extent. However this term is sometimes restricted to upward-growing free-convective cumuliform and cumulonimbiform genera to the exclusion of deep stratiform clouds. Downward-growing nimbostratus can have the same vertical extent as most large upward-growing cumulus, but its horizontal extent tends to be even greater. This sometimes leads to the exclusion of nimbostratus from the group of vertical clouds. Classifications that follow this approach usually show nimbostratus either as low-étage to denote its normal base height range, or as middle, based on the altitude range at which it normally forms. The terms *multi-level* or *multi-étage* are used for all very thick or tall cloud types including nimbostratus to avoid the association of 'vertical' with free-convective cumuliform only. Alternatively, some classifications do not recognize a vertical or multi-étage designation for any genus types and include all vertical free-convective cumuliform and cumulonimbiform genera with the low-étage clouds.

Nimbostratus and some cumulus in this group usually achieve moderate or deep vertical extent, but without towering structure. However, with sufficient airmass instability, upward-growing cumuliform clouds can grow to high towering proportions. Although genus types with vertical extent are often considered a single group, the International Civil Aviation Organization (ICAO) further distinguishes towering vertical clouds as a separate group or sub-group. It is specified that these very large cumuliform and cumulonimbiform types must be identified by their standard names or abbreviations in all aviation observations (METARS) and forecasts (TAFS) to warn pilots of possible severe weather and turbulence. When towering vertical types are considered separately, they comprise the aforementioned cumulonimbus genus and one cumulus subtype, cumulus congestus (Cu con), which is designated *towering cumulus* (Tcu) by ICAO. There is no stratiform type in this group because by definition, even very thick stratiform clouds cannot have towering vertical structure, although nimbostratus may be accompanied by embedded towering cumuliform or cumulonimbiform types.

Moderate and Deep Vertical

Moderate to deep vertical nimbostratus cloud covering the sky with a scattered layer of low stratus fractus in the middle of the upper half of the image.

- Genus nimbostratus (Ns):

This is a diffuse dark-grey non-convective stratiform layer with great horizontal extent and moderate to deep vertical development. It lacks towering structure and looks feebly illuminated from the inside. Ns normally forms from middle-étage altostratus, and develops at least moderate vertical extent when the base subsides into the low étage during precipitation that can reach moderate to heavy intensity. It commonly achieves deep vertical development when it simultaneously grows upward into the high étage due to large scale frontal or cyclonic lift. The *nimbo-* prefix refers to its ability to produce continuous rain or snow over a wide area, especially ahead of a warm front.

- Genus cumulus (Cu) – *moderate vertical extent*:

These cumuliform clouds of free convection have clear-cut medium-grey flat bases and white domed tops in the form of small sproutings and generally do not produce precipitation. They usually form in the low étage except during conditions of very low relative humidity when the clouds bases can rise into the middle altitude range.

Towering vertical cumulus congestus embedded within a layer of cumulus mediocris. Higher layer of stratocumulus stratiformis perlucidus.

These clouds are sometimes classified separately from the other vertical or multi-étage types because of their ability to produce severe turbulence.

- Genus cumulus (Cu) – *great vertical extent*:

Increasing airmass instability can cause free-convective cumulus to grow very tall to the extent that the vertical height from base to top is greater than the base-width of the cloud. The cloud base takes on a darker grey coloration and the top commonly resembles a cauliflower. This cloud type can produce moderate to heavy showers.

- Genus cumulonimbus (Cb):

This genus type is a heavy towering cumulonimbiform mass of free convective cloud with a dark-grey to nearly black base and a very high top in the form of a mountain or huge tower. Cumulonimbus can produce thunderstorms, local very heavy downpours of rain that may cause flash floods, and a variety of types of lightning including cloud-to-ground that can cause wildfires. Other convective severe weather may or may not be associated with thunderstorms and include heavy snow showers, hail, strong wind shear, downbursts, and tornadoes. Of all these possible cumulonimbus-related events, lightning is the only one of these that requires a thunderstorm to be taking place since it is the lightning that creates the thunder. Cumulonimbus clouds can form in unstable airmass conditions, but tend to be more concentrated and intense when they are associated with unstable cold fronts.

Species

Altocumulus lenticularis forming over mountains in Wyoming with lower layer of cumulus mediocris and higher layer of cirrus spissatus

Genus types are commonly divided into subtypes called *species* that indicate specific structural details which can vary according to the stability and windshear characteristics of the atmosphere at any given time and location. Despite this hierarchy, a particular species may be a subtype of more than one genus, especially if the genera are of the same physical form and are differentiated from each other mainly by altitude or étage. Some species can even be subtypes of genera that are each of different physical forms.

The species types are grouped below according to the physical forms and genera with which each is normally associated. The forms, genera, and species are listed in approximate ascending order of instability or convective activity.

Stable Stratiform Species

Of the stratiform group, high-level cirrostratus comprises two species. Cirrostratus *nebulosus* has a rather diffuse appearance lacking in structural detail. Cirrostratus *fibratus* is a species made of semi-merged filaments that are transitional to or from cirrus. Mid-level altostratus and multi-level nimbostratus always have a flat or diffuse appearance and are therefore not subdivided into species. Low-étage stratus is of the species nebulosus except when broken up into ragged sheets of stratus fractus.

Mostly Stable Cirriform Species

Cirriform clouds have three non-convective species that can form in mostly *stable* airmass conditions. Cirrus fibratus comprise filaments that may be straight, wavy, or occasionally twisted by non-convective wind shear. The species *uncinus* is similar but has upturned hooks at the ends. Cirrus *spissatus* appear as opaque patches that can show light grey shading.

Mostly Stable Stratocumuliform Species

Stratocumuliform genus-types (cirrocumulus, altocumulus, and stratocumulus) that appear in mostly stable air have two species each that can form in the high, middle, or low étages of the troposphere. The *stratiformis* species normally occur in extensive sheets or in smaller patches where there is only minimal convective activity. Clouds of the *lenticularis* species tend to have lens-like shapes tapered at the ends. They are most commonly seen as orographic mountain-wave clouds, but can occur anywhere in the troposphere where there is strong wind shear combined with sufficient airmass stability to maintain a generally flat cloud structure.

Ragged Stratiform and Cumuliform Species

The species *fractus* shows *variable* instability because it can be a subdivision of genus-types of different physical forms that have different stability characteristics. This subtype can be in the form of ragged but mostly *stable* stratiform sheets (stratus fractus) or small ragged cumuliform heaps with somewhat greater instability (cumulus fractus). When they form at low altitudes, stratiform and cumuliform genus-types can be torn up into shreds by brisk low level winds that create mechanical turbulance against the ground. Fractus clouds can form in precipitation at low altitudes, with or without brisk or gusty winds. They are closely associated with precipitating cloud systems of considerable vertical and sometimes horizontal extent, so they are also classified as *accessory clouds* under the name *pannus*.

Partly Unstable Cirriform, Stratocumuliform, and Cumuliform Species

These species are subdivisions of genus types that occur in partly unstable air. The species *castellanus* appears when a mostly stable stratocumuliform or cirriform layer becomes disturbed by localized areas of airmass instability. Castellanus resembles the turrets of a castle when viewed from the side, and can be found with stratocumuliform genera at any tropospheric altitude level and with limited-convective patches of high-étage cirrus. Clouds of the more detached tufted *floccus* species are subdivisions of genus-types which may be cirriform or cumuliform. They are sometimes seen with the genera cirrus, cirrocumulus, and altocumulus. However tufted floccus clouds are not generally found in the low étage.

Mostly Unstable Cumuliform Species

More general airmass instability in the troposphere tends to produce clouds of the more freely convective cumulus genus type, whose species are mainly indicators of degrees of atmospheric instability and resultant vertical development of the clouds. A cumulus cloud initially forms as a cloudlet of the species *humilis* that shows only slight vertical development. If the air becomes more unstable, the cloud tends to grow vertically into the species *mediocris*, then *congestus*, the tallest cumulus species.

Unstable Cumulonimbiform Species

With highly unstable atmospheric conditions, large cumulus may continue to grow into cumulo-nimbus *calvus* (essentially a very tall congestus cloud that produces thunder), then ultimately into the species *capillatus* when supercooled water droplets at the top turn into ice crystals giving it a cirriform appearance.

Varieties

Genus and species types are further subdivided into *varieties* whose names can appear after the species name to provide a fuller description of a cloud. Some cloud varieties are not restricted to a specific étage or form, and can therefore be common to more than one genus or species.

Opacity-based

All cloud varieties fall into one of two main groups. One group identifies the opacities of particular low and middle étage cloud structures and comprises the varieties *translucidus* (thin translucent), *perlucidus* (thick opaque with translucent breaks), and *opacus* (thick opaque). These varieties are always identifiable for cloud genera and species with variable opacity. All three are associated with the stratiformis species of altocumulus and stratocumulus. However, only two varieties are seen with altostratus and stratus nebulosus whose uniform structures prevent the formation of a perlucidus variety. Opacity-based varieties are not applied to high-étage clouds because they are always translucent, or in the case of cirrus spissatus, always opaque. Similarly, these varieties are also not associated with moderate and towering vertical clouds because they are always opaque.

Pattern-based

Cirrus fibratus radiatus over ESO's La Silla Observatory

A second group describes the occasional arrangements of cloud structures into particular patterns that are discernible by a surface-based observer (cloud fields usually being visible only from a significant altitude above the formations). These varieties are not always present with the genera and species with which they are otherwise associated, but only appear when atmospheric conditions favor their formation. *Intortus* and *vertebratus* varieties occur on occasion with cirrus fibratus. They are respectively filaments twisted into irregular shapes, and those that are arranged in fishbone patterns, usually by uneven wind currents that favor the formation of these varieties. The variety *radiatus* is associated with cloud rows of a particular type that appear to converge at the horizon. It is sometimes seen with the fibratus and uncinus species of cirrus, the stratiformis species of altocumulus and stratocumulus, the mediocris and sometimes humilis species of cumulus, and with the genus altostratus.

Another variety, *duplicatus* (closely spaced layers of the same type, one above the other), is sometimes found with cirrus of both the fibratus and uncinus species, and with altocumulus and stratocumulus of the species stratiformis and lenticularis. The variety *undulatus* (having a wavy undulating base) can occur with any clouds of the species stratiformis or lenticularis, and with altostratus. It is only rarely observed with stratus nebulosus. The variety *lacunosus* is caused by localized downdrafts that create circular holes in the form of a honeycomb or net. It is occasionally seen with cirrocumulus and altocumulus of the species stratiformis, castellanus, and floccus, and with stratocumulus of the species stratiformis and castellanus.

Combinations

It is possible for some species to show combined varieties at one time, especially if one variety is opacity-based and the other is pattern-based. An example of this would be a layer of altocumulus stratiformis arranged in seemingly converging rows separated by small breaks. The full technical name of a cloud in this configuration would be *altocumulus stratiformis radiatus perlucidus*, which would identify respectively its genus, species, and two combined varieties.

Accessory Clouds, Supplementary Features, and Other Derivative Formations

Cumulus and stratocumulus made orange by the sun rising

Supplementary features and accessory clouds are not further subdivisions of cloud types below the species and variety level. Rather, they are either *hydrometeors* or special cloud formations with their own Latin names that form in association with certain cloud genera, species, and varieties. Supplementary features, whether in the form of clouds or precipitation, are directly attached to the main genus-cloud. Accessory clouds, by contrast, are generally detached from the main cloud.

Precipitation-based Supplementary Features

One group of supplementary features are not actual cloud formations, but precipitation that falls when water droplets or ice crystals that make up visible clouds have grown too heavy to remain aloft. *Virga* is a feature seen with clouds producing precipitation that evaporates before reaching the ground, these being of the genera cirrocumulus, altocumulus, altostratus, nimbostratus, stratocumulus, cumulus, and cumulonimbus.

When the precipitation reaches the ground without completely evaporating, it is designated as the feature *praecipitatio*. This normally occurs with altostratus opacus, which can produce widespread but usually light precipitation, and with thicker clouds that show significant vertical development. Of the latter, *upward-growing* cumulus mediocris produces only isolated light showers, while *downward growing* nimbostratus is capable of heavier, more extensive precipitation. Towering vertical clouds have the greatest ability to produce intense precipitation events, but these tend to be localized unless organized along fast-moving cold fronts. Showers of moderate to heavy intensity can fall from cumulus congestus clouds. Cumulonimbus, the largest of all cloud genera, has the capacity to produce very heavy showers. Low stratus clouds usually produce only light precipitation, but this always occurs as the feature praecipitatio due to the fact this cloud genus lies too close to the ground to allow for the formation of virga.

Cumulonimbus dissipating at dusk

Cloud-based Supplementary Features

Incus is the most type-specific supplementary feature, seen only with cumulonimbus of the species capillatus. A cumulonimbus incus cloud top is one that has spread out into a clear anvil shape as a result of rising air currents hitting the stability layer at the tropopause where the air no longer continues to get colder with increasing altitude.

The *mamma* feature forms on the bases of clouds as downward-facing bubble-like protuberances caused by localized downdrafts within the cloud. It is also sometimes called *mammatus*, an earlier version of the term used before a standardization of Latin nomenclature brought about by the World Meterorological Organization during the 20th century. The best-known is cumulonimbus with mammatus, but the mamma feature is also seen occasionally with cirrus, cirrocumulus, altocumulus, altostratus, and stratocumulus.

A *tuba* feature is a cloud column that may hang from the bottom of a cumulus or cumulonimbus. A newly formed or poorly organized column might be comparatively benign, but can quickly intensify into a funnel cloud or tornado.

An *arcus* feature is a roll cloud with ragged edges attached to the lower front part of cumulus congestus or cumulonimbus that forms along the leading edge of a squall line or thunderstorm outflow. A large arcus formation can have the appearance of a dark menacing arch.

There are some arcus-like clouds that form as a consequence of interactions with specific geographical features rather than with a parent cloud. Perhaps the strangest geographically specific cloud of this type is the Morning Glory, a rolling cylindrical cloud that appears unpredictably over the Gulf of Carpentaria in Northern Australia. Associated with a powerful "ripple" in the atmosphere, the cloud may be "surfed" in glider aircraft. It has been officially suggested that roll clouds of this type that are not attached to a parent cloud be reclassified as a new species of stratocumulus, possibly with the Latin name *volutus*.

Accessory Clouds

Supplementary cloud formations detached from the main cloud are known as accessory clouds. The heavier precipitating clouds, nimbostratus, towering cumulus (cumulus congestus), and cumulonimbus typically see the formation in precipitation of the *pannus* feature, low ragged clouds of the genera and species cumulus fractus or stratus fractus.

After the pannus types, the remaining accessory clouds comprise formations that are associated mainly with upward-growing cumuliform and cumulonimbiform clouds of free convection. *Pileus* is a cap cloud that can form over a cumulonimbus or large cumulus cloud, whereas a *velum* feature is a thin horizontal sheet that sometimes forms like an apron around the middle or in front of the parent cloud.

Under conditions of strong atmospheric wind shear and instability, wave-like undulatus formations may break into regularly spaced crests. This variant has no separate WMO Latin designation, but is sometimes known informally as a Kelvin–Helmholtz (wave) cloud. This phenomenon has also been observed in cloud formations over other planets and even in the sun's atmosphere. It has been formally suggested that this wave cloud be classified as a supplementary feature, possibly with the Latin name *fluctus*. Another wave-like cloud feature that is distinct from the variety undulatus has been given the Latin name *asperatus*. It has been recommended for formal classification as a supplementary feature using its suggested Latin name.

A circular fall-streak hole occasionally forms in a thin layer of supercooled altocumulus or cirrocumulus. Fall streaks consisting of virga or wisps of cirrus are usually seen beneath the hole as ice crystals fall out to a lower altitude. This type of hole is usually larger than typical lacunosus holes, and a formal recommendation has been made to classify it as a supplementary feature, possibly with the Latin name *cavus*.

Mother Clouds

Clouds initially form in clear air or become clouds when fog rises above surface level. The genus of a newly formed cloud is determined mainly by air mass characteristics such as stability and moisture content. If these characteristics change over time, the genus tends to change accordingly.

When this happens, the original genus is called a *mother cloud*. If the mother cloud retains much of its original form after the appearance of the new genus, it is termed a *genitus* cloud. One example of this is *stratocumulus cumulogenitus*, a stratocumulus cloud formed by the partial spreading of a cumulus type when there is a loss of convective lift. If the mother cloud undergoes a complete change in genus, it is considered to be a *mutatus* cloud.

Cumulus partly spreading into stratocumulus cumulogenitus over the port of Piraeus in Greece

It has been officially recommended that the genitus category be expanded to include certain types that do not originate from pre-existing clouds or as the result of any natural atmospheric processes. Among vertically developed clouds, these may include *flammagenitus* for cumulus congestus or cumulonimbus that are formed by large scale fires or volcanic eruptions. Smaller low-étage "pyrocumulus" or "fumulus" clouds formed by contained industrial activity could be classified as cumulus *homogenitus*. Contrails formed from the exhaust of aircraft flying in the high étage can persist and spread into formations resembling any of the high cloud genus-types. These variants have no special WMO designations, but are sometimes given the faux-Latin name Aviaticus. Persistent contrails have been identified as candidates for possible inclusion in the genitus category as cirrus, cirrostratus, or cirrocumulus homogenitus

Stratocumulus Fields

Stratocumulus clouds can be organized into "fields" that take on certain specially classified shapes and characteristics. In general, these fields are more discernible from high altitudes than from ground level. They can often be found in the following forms:

- Actinoform, which resembles a leaf or a spoked wheel.

- Closed cell, which is cloudy in the center and clear on the edges, similar to a filled honeycomb.

- Open cell, which resembles an empty honeycomb, with clouds around the edges and clear, open space in the middle.

Vortex Streets

Cirrus fibratus intortus formed into a Kármán vortex street at twilight

These patterns are formed from a phenomenon known as a Kármán vortex which is named after the engineer and fluid dynamicist Theodore von Kármán. When wind driven clouds are forced through a mountain range, or when ocean wind driven clouds encounter a high elevation island, they can begin to circle the mountain or high land mass. They can form at any altitude in the troposphere and are not restricted to any particular cloud type.

Formation and Distribution

How The Air Becomes Saturated

Cloud evolution in under a minute.

Late-summer rainstorm in Denmark. Nearly black color of base indicates main cloud in foreground probably cumulonimbus.

Air can become saturated as a result of being cooled to its dew point or by having moisture added from an adjacent source. Adiabatic cooling occurs when one or more of three possible lifting agents - cyclonic/frontal, convective, or orographic — causes air containing invisible water vapor to rise and cool to its dew point, the temperature at which the air becomes saturated. The main mechanism behind this process is adiabatic cooling. If the air is cooled to its dew point and becomes saturated, it normally sheds vapor it can no longer retain, which condenses into cloud. Water vapor in saturated air is normally attracted to condensation nuclei such as dust and salt particles that are small enough to be held aloft by normal circulation of the air.

Frontal and cyclonic lift occur when stable air is forced aloft at weather fronts and around centers of low pressure. Warm fronts associated with extratropical cyclones tend to generate mostly cirriform and stratiform clouds over a wide area unless the approaching warm airmass is unstable, in which case cumulus congestus or cumulonimbus clouds will usually be embedded in the main precipitating cloud layer. Cold fronts are usually faster moving and generate a narrower line of clouds which are mostly stratocumuliform, cumuliform, or cumulonimbiform depending on the stability of the warm air mass just ahead of the front.

Another agent is the convective upward motion of air caused by daytime solar heating at surface level. Airmass instability allows for the formation of cumuliform clouds that can produce showers if the air is sufficiently moist. On comparatively rare occasions, convective lift can be powerful enough to penetrate the tropopause and push the cloud top into the stratosphere.

A third source of lift is wind circulation forcing air over a physical barrier such as a mountain (orographic lift). If the air is generally stable, nothing more than lenticular cap clouds will form. However, if the air becomes sufficiently moist and unstable, orographic showers or thunderstorms may appear.

Windy evening twilight enhanced by the Sun's angle, can visually mimic a tornado resulting from orographic lift

Along with adiabatic cooling that requires a lifting agent, there are three major non-adiabatic mechanisms for lowering the temperature of the air to its dew point. Conductive, radiational, and evaporative cooling require no lifting mechanism and can cause condensation at surface level resulting in the formation of fog.

There are several main sources of water vapor that can be added to the air as a way of achieving saturation without any cooling process: Water or moist ground, precipitation or virga, and transpiration from plants

Convergence Along Low-Pressure Zones

Although the local distribution of clouds can be significantly influenced by topography, the global prevalence of cloud cover tends to vary more by latitude. It is most prevalent globally in and along low pressure zones of surface atmospheric convergence which encircle the Earth close to the equator and near the 50th parallels of latitude in the northern and southern hemispheres. The adiabatic cooling processes that lead to the creation of clouds by way of lifting agents are all associated with convergence; a process that involves the horizontal inflow and accumulation of air at a given location, as well as the rate at which this happens. Near the equator, increased cloudiness is due to the presence of the low-pressure Intertropical Convergence Zone (ITCZ) where very warm and unstable air promotes mostly cumuliform and cumulonimbiform clouds. Clouds of virtually any type can form along the mid-latitude convergence zones depending on the stability and moisture content of the air. These extratropical convergence zones are occupied by the polar fronts where air masses of polar origin meet and clash with those of tropical or subtropical origin. This leads to the formation of weather-making extratropical cyclones composed of cloud systems that may be stable or unstable to varying degrees according to the stability characteristics of the various airmasses that are in conflict.

Divergence Along High Pressure Zones

Divergence is the opposite of convergence. In the Earth's atmosphere, it involves the horizontal outflow of air from the upper part of a rising column of air, or from the lower part of a subsiding column often associated with an area or ridge of high pressure. Cloudiness tends to be least prevalent near the poles and in the subtropics close to the 20th parallels, north and south. The latter are sometimes referred to as the horse latitudes. The presence of a large-scale high-pressure subtropical ridge on each side of the equator reduces cloudiness at these low latitudes. Similar patterns also occur at higher latitudes in both hemispheres.

Polar Stratospheric

Stratospheric nacreous clouds over Antarctica

Polar stratospheric clouds show little variation in structure and are limited to a single very high range of altitude of about 15,000–25,000 m (49,200–82,000 ft), so they are not classified into étages, genus types, species, or varieties in the manner of tropospheric clouds. Instead, the classification is alpha-numeric and is based on chemical makeup rather than variations in physical appearance.

Types and Subtypes

Nacreous and Non-nacreous (Very High Cirriform)

- Type 1 (Non-nacreous): This type contains frozen or supercooled nitric acid and water droplets and lacks any special coloration. It is dividable into subtype 1A which is mostly made up of water ice crystals and frozen nitric acid, 1B which consists of supercooled droplets of nitric and sulfuric acid, and subtype 1C which comprises small particles of nitric acid. Nacreous type 2 is sometimes associated or embedded.

- Type 2 (Nacreous): Nacreous polar stratospheric cloud consists of ice crystals only and generally shows mother-of-pearl colors.

Formation and distribution

Polar stratospheric clouds form in the lowest part of the stratosphere during the winter, at the altitude and during the season that produces the coldest temperatures and therefore the best chances of triggering condensation caused by adiabatic cooling. They are typically very thin with an undulating cirriform appearance. Moisture is scarce in the stratosphere, so nacreous and non-nacreous cloud at this altitude range is rare and is usually restricted to polar regions in the winter where the air is coldest.

Polar Mesospheric

Noctilucent cloud over Estonia

Polar mesospheric clouds form at a single extreme altitude range of about 80 to 85 km (50 to 53 mi) and are consequently not classified into more than one étage. They are given the Latin name noctilucent because of their illumination well after sunset and before sunrise. An alpha-numeric classification is used to identify variations in physical appearance.

Types and Subtypes

- Type 1: The first type is characterized by very tenuous filaments resembling cirrus fibratus.

- Type 2: This type comprises bands in the form of long streaks, often in groups or interwoven at small angles, similar to cirrus intortus. It is dividable into two subtypes; 2A where the streaks have diffuse, blurred edges, and 2B where they have sharply defined edges.

- Type 3: Billows in the form of short streaks can be seen that are clearly spaced and roughly parallel. Subtype 3A has short, straight, narrow streaks while 3B has wave-like streaks similar to cirrus undulatus.

- Type 4: This shows whirls in the form of partial or rarely complete rings with dark centers. With subtype 4A, the whirls are of small angular radius and have a similar appearance to surface water ripples. 4B is characterized by simple curves of medium angular radius with one or more bands. Subtype 4C has whirls with large-scale ring structure.

Formation and Distribution

Polar mesospheric clouds are the highest in the atmosphere and form near the top of the mesosphere at about ten times the altitude of tropospheric high clouds. From ground level, they can occasionally be seen illuminated by the sun during deep twilight. Ongoing research indicates that convective lift in the mesosphere is strong enough during the polar summer to cause adiabatic cooling of small amount of water vapour to the point of saturation. This tends to produce the coldest temperatures in the entire atmosphere just below the mesopause. These conditions result in the best environment for the formation of polar mesospheric clouds. There is also evidence that smoke particles from burnt-up meteors provide much of the condensation nuclei required for the formation of noctilucent cloud.

Distribution in the mesosphere is similar to the stratosphere except at much higher altitudes. Because of the need for maximum cooling of the water vapor to produce noctilucent clouds, their distribution tends to be restricted to polar regions of Earth. A major seasonal difference is that convective lift from below the mesosphere pushes very scarce water vapor to higher colder altitudes required for cloud formation during the respective summer seasons in the northern and southern hemispheres. Sightings are rare more than 45 degrees south of the north pole or north of the south pole.

Throughout The Homosphere

Luminance and Reflectivity

The luminance or brightness of a cloud in the homosphere (which includes the troposphere, stratosphere, and mesosphere) is determined by how light is reflected, scattered, and transmitted by the cloud's particles. Its brightness may also be affected by the presence of haze or photometeors such as halos and rainbows. In the troposphere, dense, deep clouds exhibit a high reflectance (70% to 95%) throughout the visible spectrum. Tiny particles of water are densely packed and sunlight cannot penetrate far into the cloud before it is reflected out, giving a cloud its characteristic white color, especially when viewed from the top. Cloud droplets tend to scatter light efficiently, so that the intensity of the solar radiation decreases with depth into the gases. As a result, the cloud base can vary from a very light to very-dark-grey depending on the cloud's thickness and how much light is being reflected or transmitted back to the observer. High thin tropospheric clouds reflect less light because of the comparatively low concentration of constituent ice crystals or supercooled water droplets which results in a slightly off-white appearance. However, a thick dense ice-crystal cloud appears brilliant white with pronounced grey shading because of its greater reflectivity.

As a tropospheric cloud matures, the dense water droplets may combine to produce larger droplets. If the droplets become too large and heavy to be kept aloft by the air circulation, they will fall from the cloud as rain. By this process of accumulation, the space between droplets becomes increasingly larger, permitting light to penetrate farther into the cloud. If the cloud is sufficiently large and the droplets within are spaced far enough apart, a percentage of the light that enters the cloud is not reflected back out but is absorbed giving the cloud a darker look. A simple example of this is one's being able to see farther in heavy rain than in heavy fog. This process of reflection/absorption is what causes the range of cloud color from white to black.

Coloration

An occurrence of altocumulus and cirrocumulus cloud iridescence

Effect of sunlight before sunset. Bangalore, India.

Striking cloud colorations can be seen at many altitudes in the homosphere. The color of a cloud is usually the same as the incident light.

During daytime when the sun is relatively high in the sky, tropospheric clouds generally appear bright white on top with varying shades of grey underneath. Thin clouds may look white or appear to have acquired the color of their environment or background. Red, orange, and pink clouds occur almost entirely at sunrise/sunset and are the result of the scattering of sunlight by the atmosphere. When the sun is just below the horizon, low-etage clouds are gray, middle clouds appear rose-colored, and high-etage clouds are white or off-white. Clouds at night are black or dark grey in a moonless sky, or whitish when illuminated by the moon. They may also reflect the colors of large fires, city lights, or auroras that might be present.

A cumulonimbus cloud that appears to have a greenish/bluish tint is a sign that it contains extremely high amounts of water; hail or rain which scatter light in a way that gives the cloud a blue color. A green colorization occurs mostly late in the day when the sun is comparatively low in the sky and the incident sunlight has a reddish tinge that appears green when illuminating a very tall bluish cloud. Supercell type storms are more likely to be characterized by this but any storm can appear this way. Coloration such as this does not directly indicate that it is a severe thunderstorm, it only confirms its potential. Since a green/blue tint signifies copious amounts of water, a strong updraft to support it, high winds from the storm raining out, and wet hail; all elements that improve the chance for it to become severe, can all be inferred from this. In addition, the stronger the updraft is, the more likely the storm is to undergo tornadogenesis and to produce large hail and high winds.

Yellowish clouds may be seen in the troposphere in the late spring through early fall months during forest fire season. The yellow color is due to the presence of pollutants in the smoke. Yellowish clouds caused by the presence of nitrogen dioxide are sometimes seen in urban areas with high air pollution levels.

Particles in the atmosphere and the sun's angle enhance cloud colors at evening twilight

In high latitude regions of the stratosphere, nacreous clouds occasionally found there during the polar winter tend to display quite striking displays of mother-of-pearl colorations. This is due to the refraction and diffusion of the sun's rays through thin clouds with supercooled droplets that often contain compounds other than water. At still higher altitudes up in the mesosphere, noctilucent clouds made of ice crystals are sometimes seen in polar regions in the summer. They typically have a bluish or silvery white coloration that can resemble brightly illuminated cirrus. Noctilucent clouds may occasionally take on more of a red or orange hue.

Effects on Climate and The Atmosphere

Global cloud cover, averaged over the month of October 2009. NASA composite satellite image; larger image available here:

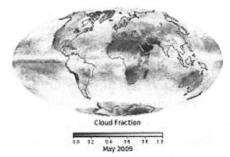

These maps display the fraction of Earth's area that was cloudy on average during each month from January 2005 to August 2013. The measurements were collected by the Moderate Resolution Imaging Spectroradiometer (MODIS) on NASA's Terra satellite. Colors range from blue (no clouds) to white (totally cloudy). Like a digital camera, MODIS collects information in gridded boxes, or pixels. Cloud fraction is the portion of each pixel that is covered by clouds. Colors range from blue (no clouds) to white (totally cloudy).

The role of tropospheric clouds in regulating weather and climate remains a leading source of uncertainty in projections of global warming. This uncertainty arises because of the delicate balance of processes related to clouds, spanning scales from millimeters to planetary. Hence, interactions between the large-scale (synoptic meteorology) and clouds becomes difficult to represent in global models.

The complexity and diversity of clouds, as outlined above, adds to the problem. On the one hand, white-colored cloud tops promote cooling of Earth's surface by reflecting *short-wave* radiation from the sun. Most of the sunlight that reaches the ground is absorbed, warming the surface, which emits radiation upward at longer, *infrared*, wavelengths. At these wavelengths, however, water in the clouds acts as an efficient absorber. The water reacts by radiating, also in the infrared, both upward and downward, and the downward *long-wave* radiation results in some warming at the surface. This is analogous to the greenhouse effect of greenhouse gases and water vapor.

High-étage tropospheric genus-types, *cirrus*, *cirrocumulus*, and *cirrostratus*, particularly show this duality with both short-wave albedo cooling and long-wave greenhouse warming effects. On the whole though, *ice-crystal* clouds in the upper troposphere tend to favor net *warming*. However, the *cooling* effect is dominant with lower clouds made of very small *water droplets*, especially when they form in extensive sheets that block out more of the sun. These include middle-étage layers of *altocumulus* and *altostratus* as well as low *stratocumulus*, and *stratus* that have droplets with an average radius of about 0.002 mm (0.00008 in). Small-droplet aerosols are not good at absorbing long-wave radiation reflected back from Earth, so there is a net cooling with almost no long-wave effect. This effect is particularly pronounced with low-étage clouds that form over water.

Low and vertical heaps of *cumulus, towering cumulus,* and *cumulonimbus* are made of larger water droplets ranging in radius from 0.005 to about 0.015 mm. Nimbostratus cloud droplets can also be quite large, up to 0.015 mm radius. These larger droplets associated with vertically developed clouds are better able to trap the long-wave radiation thus mitigating the cooling effect to some degree. However, these large often precipitating clouds are variable or unpredictable in their overall effect because of variations in their concentration, distribution, and vertical extent. Measurements taken by NASA indicate that on the whole, the effects of low and middle étage clouds that tend to promote cooling are outweighing the warming effects of high layers and the variable outcomes associated with multi-étage or vertically developed clouds.

As difficult as it is to evaluate the effects of current cloud cover characteristics on climate change, it is even more problematic to predict the outcome of this change with respect to future cloud patterns and events. As a consequence, much research has focused on the response of low and vertical clouds to a changing climate. Leading global models can produce quite different results, however, with some showing increasing low-étage clouds and others showing decreases.

In the stratosphere, Type I non-nacreous clouds are known to have harmful effects over the polar regions of Earth. They become catalysts which convert relatively benign man-made chlorine into active free radicals like chlorine monoxide which are destructive of the stratospheric ozone layer.

Polar mesospheric clouds are not common or widespread enough to have a significant effect on climate. However, an increasing frequency of occurrence of noctilucent clouds since the 19th century may be the result of climate change.

Global Brightening

New research indicates a global brightening trend. The details are not fully understood, but much of the global dimming (and subsequent reversal) is thought to be a consequence of changes in aerosol loading in the atmosphere, especially sulfur-based aerosol associated with biomass burning and urban pollution. Changes in aerosol burden can have indirect effects on clouds by changing the droplet size distribution, or the lifetime and precipitation characteristics of clouds.

Extraterrestrial

Cloud cover has been seen on most other planets in the solar system. Venus's thick clouds are composed of sulfur dioxide and appear to be almost entirely stratiform. They are arranged in three main layers at altitudes of 45 to 65 km that obscure the planet's surface and can produce virga. No embedded cumuliform types have been identified, but broken stratocumuliform wave formations are sometimes seen in the top layer that reveal more continuous layer clouds underneath. On Mars, noctilucent, cirrus, cirrocumulus and stratocumulus composed of water-ice have been detected mostly near the poles. Water-ice fogs have also been detected on this planet.

Both Jupiter and Saturn have an outer cirriform cloud deck composed of ammonia, an intermediate stratiform haze-cloud layer made of ammonium hydrosulfide, and an inner deck of cumulus water clouds. Embedded cumulonimbus are known to exist near the Great Red Spot on Jupiter. The same category-types can be found covering Uranus, and Neptune, but are all composed of methane. Saturn's moon Titan has cirrus clouds believed to be composed largely of methane. The

Cassini–Huygens Saturn mission uncovered evidence of a fluid cycle on Titan, including lakes near the poles and fluvial channels on the surface of the moon.

Some planets outside the solar system are known to have atmospheric clouds. In October 2013, the detection of high altitude optically thick clouds in the atmosphere of exoplanet Kepler-7b was announced, and, in December 2013, also in the atmospheres of GJ 436 b and GJ 1214 b.

Rain

Rain is liquid water in the form of droplets that have condensed from atmospheric water vapor and then precipitated—that is, become heavy enough to fall under gravity. Rain is a major component of the water cycle and is responsible for depositing most of the fresh water on the Earth. It provides suitable conditions for many types of ecosystems, as well as water for hydroelectric power plants and crop irrigation.

The major cause of rain production is moisture moving along three-dimensional zones of temperature and moisture contrasts known as weather fronts. If enough moisture and upward motion is present, precipitation falls from convective clouds (those with strong upward vertical motion) such as cumulonimbus (thunder clouds) which can organize into narrow rainbands. In mountainous areas, heavy precipitation is possible where upslope flow is maximized within windward sides of the terrain at elevation which forces moist air to condense and fall out as rainfall along the sides of mountains. On the leeward side of mountains, desert climates can exist due to the dry air caused by downslope flow which causes heating and drying of the air mass. The movement of the monsoon trough, or intertropical convergence zone, brings rainy seasons to savannah climes.

The urban heat island effect leads to increased rainfall, both in amounts and intensity, downwind of cities. Global warming is also causing changes in the precipitation pattern globally, including wetter conditions across eastern North America and drier conditions in the tropics. Antarctica is the driest continent. The globally averaged annual precipitation over land is 715 mm (28.1 in), but over the whole Earth it is much higher at 990 mm (39 in). Climate classification systems such as the Köppen climate classification system use average annual rainfall to help differentiate between differing climate regimes. Rainfall is measured using rain gauges. Rainfall amounts can be estimated by weather radar.

Rain is also known or suspected on other planets, where it may be composed of methane, neon, sulfuric acid, or even iron rather than water.

Formation

Water-saturated Air

Air contains water vapor, and the amount of water in a given mass of dry air, known as the *mixing ratio*, is measured in grams of water per kilogram of dry air (g/kg). The amount of moisture in air is also commonly reported as relative humidity; which is the percentage of the total water vapor air can hold at a particular air temperature. How much water vapor a parcel of air can contain before it becomes saturated (100% relative humidity) and forms into a cloud (a group of visible and tiny water and ice particles suspended above the Earth's surface) depends on its temperature. Warmer air can contain more water vapor than cooler air before becoming saturated. Therefore, one way to saturate a parcel of air is to cool it. The dew point is the temperature to which a parcel must be cooled in order to become saturated.

There are four main mechanisms for cooling the air to its dew point: adiabatic cooling, conductive cooling, radiational cooling, and evaporative cooling. Adiabatic cooling occurs when air rises and expands. The air can rise due to convection, large-scale atmospheric motions, or a physical barrier such as a mountain (orographic lift). Conductive cooling occurs when the air comes into contact with a colder surface, usually by being blown from one surface to another, for example from a liquid water surface to colder land. Radiational cooling occurs due to the emission of infrared radiation, either by the air or by the surface underneath. Evaporative cooling occurs when moisture is added to the air through evaporation, which forces the air temperature to cool to its wet-bulb temperature, or until it reaches saturation.

The main ways water vapor is added to the air are: wind convergence into areas of upward motion, precipitation or virga falling from above, daytime heating evaporating water from the surface of oceans, water bodies or wet land, transpiration from plants, cool or dry air moving over warmer water, and lifting air over mountains. Water vapor normally begins to condense on condensation nuclei such as dust, ice, and salt in order to form clouds. Elevated portions of weather fronts (which are three-dimensional in nature) force broad areas of upward motion within the Earth's atmosphere which form clouds decks such as altostratus or cirrostratus. Stratus is a stable cloud deck which tends to form when a cool, stable air mass is trapped underneath a warm air mass. It can also form due to the lifting of advection fog during breezy conditions.

Coalescence and Fragmentation

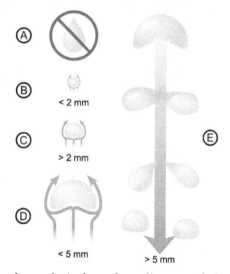

The shape of rain drops depending upon their size

Coalescence occurs when water droplets fuse to create larger water droplets. Air resistance typically causes the water droplets in a cloud to remain stationary. When air turbulence occurs, water droplets collide, producing larger droplets. As these larger water droplets descend, coalescence continues, so that drops become heavy enough to overcome air resistance and fall as rain. Coalescence generally happens most often in clouds above freezing, and is also known as the warm rain process. In clouds below freezing, when ice crystals gain enough mass they begin to fall. This generally requires more mass than coalescence when occurring between the crystal and neighboring water droplets. This process is temperature dependent, as supercooled water droplets only exist in a cloud that is below freezing. In addition, because of the great temperature difference between cloud and ground level, these ice crystals may melt as they fall and become rain.

Raindrops have sizes ranging from 0.1 to 9 mm (0.0039 to 0.3543 in) mean diameter, above which they tend to break up. Smaller drops are called cloud droplets, and their shape is spherical. As a raindrop increases in size, its shape becomes more oblate, with its largest cross-section facing the oncoming airflow. Large rain drops become increasingly flattened on the bottom, like hamburger buns; very large ones are shaped like parachutes. Contrary to popular belief, their shape does not resemble a teardrop. The biggest raindrops on Earth were recorded over Brazil and the Marshall Islands in 2004 — some of them were as large as 10 mm (0.39 in). The large size is explained by condensation on large smoke particles or by collisions between drops in small regions with particularly high content of liquid water.

Rain drops associated with melting hail tend to be larger than other rain drops.

A raindrop on a leaf

Intensity and duration of rainfall are usually inversely related, i.e., high intensity storms are likely to be of short duration and low intensity storms can have a long duration.

Droplet Size Distribution

The final droplet size distribution is an exponential distribution. The number of droplets with diameter between d and $D + dD$ per unit volume of space is $n(d) = n_0 e^{-d/\langle d \rangle} dD$. This is commonly referred to as the Marshall–Palmer law after the researchers who first characterized it. The parameters are somewhat temperature-dependent, and the slope also scales with the rate of rainfall $\langle d \rangle^{-1} = 41R^{-0.21}$ (d in centimeters and R in millimetres per hour).

Deviations can occur for small droplets and during different rainfall conditions. The distribution tends to fit averaged rainfall, while instantaneous size spectra often deviate and have been modeled as gamma distributions. The distribution has an upper limit due to droplet fragmentation.

Raindrop Impacts

Raindrops impact at their terminal velocity, which is greater for larger drops due to their larger mass to drag ratio. At sea level and without wind, 0.5 mm (0.020 in) drizzle impacts at 2 m/s (6.6 ft/s) or 7.2 km/h (4.5 mph), while large 5 mm (0.20 in) drops impact at around 9 m/s (30 ft/s) or 32 km/h (20 mph).

Rain falling on loosely packed material such as newly fallen ash can produce dimples that can be fossilized. The air density dependence of the maximum raindrop diameter together with fossil raindrop imprints has been used to constrain the density of the air 2.7 billion years ago.

The sound of raindrops hitting water is caused by bubbles of air oscillating underwater.

The METAR code for rain is RA, while the coding for rain showers is SHRA.

Virga

In certain conditions precipitation may fall from a cloud but then evaporates or sublimes before reaching the ground. This is termed virga and is more often seen in hot and dry climates.

Causes

Frontal Activity

Stratiform (a broad shield of precipitation with a relatively similar intensity) and dynamic precipitation (convective precipitation which is showery in nature with large changes in intensity over short distances) occur as a consequence of slow ascent of air in synoptic systems (on the order of cm/s), such as in the vicinity of cold fronts and near and poleward of surface warm fronts. Similar ascent is seen around tropical cyclones outside of the eyewall, and in comma-head precipitation patterns around mid-latitude cyclones. A wide variety of weather can be found along an occluded front, with thunderstorms possible, but usually their passage is associated with a drying of the air mass. Occluded fronts usually form around mature low-pressure areas. What separates rainfall from other precipitation types, such as ice pellets and snow, is the presence of a thick layer of air aloft which is above the melting point of water, which melts the frozen precipitation well before it reaches the ground. If there is a shallow near surface layer that is below freezing, freezing rain (rain which freezes on contact with surfaces in subfreezing environments) will result. Hail becomes an increasingly infrequent occurrence when the freezing level within the atmosphere exceeds 3,400 m (11,000 ft) above ground level.

Convection

Convective rain, or showery precipitation, occurs from convective clouds (e.g., cumulonimbus or cumulus congestus). It falls as showers with rapidly changing intensity. Convective precipitation falls over a certain area for a relatively short time, as convective clouds have limited horizontal extent. Most precipitation in the tropics appears to be convective; however, it has been suggested that stratiform precipitation also occurs. Graupel and hail indicate convection. In mid-latitudes, convective precipitation is intermittent and often associated with baroclinic boundaries such as cold fronts, squall lines, and warm fronts.

Convective precipitation

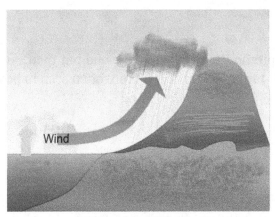

Orographic precipitation

Orographic precipitation occurs on the windward side of mountains and is caused by the rising air motion of a large-scale flow of moist air across the mountain ridge, resulting in adiabatic cooling and condensation. In mountainous parts of the world subjected to relatively consistent winds (for example, the trade winds), a more moist climate usually prevails on the windward side of a mountain than on the leeward or downwind side. Moisture is removed by orographic lift, leaving drier air on the descending and generally warming, leeward side where a rain shadow is observed.

In Hawaii, Mount Wai'ale'ale, on the island of Kauai, is notable for its extreme rainfall, as it has the second highest average annual rainfall on Earth, with 12,000 mm (460 in). Systems known as Kona storms affect the state with heavy rains between October and April. Local climates vary considerably on each island due to their topography, divisible into windward (*Ko'olau*) and leeward (*Kona*) regions based upon location relative to the higher mountains. Windward sides face the east to northeast trade winds and receive much more rainfall; leeward sides are drier and sunnier, with less rain and less cloud cover.

In South America, the Andes mountain range blocks Pacific moisture that arrives in that continent, resulting in a desertlike climate just downwind across western Argentina. The Sierra Nevada range creates the same effect in North America forming the Great Basin and Mojave Deserts.

Within The Tropics

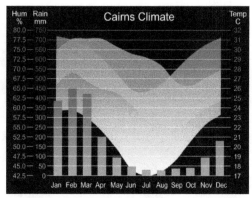

Rainfall distribution by month in Cairns showing the extent of the wet season at that location

The wet, or rainy, season is the time of year, covering one or more months, when most of the average annual rainfall in a region falls. The term *green season* is also sometimes used as a euphemism by tourist authorities. Areas with wet seasons are dispersed across portions of the tropics and subtropics. Savanna climates and areas with monsoon regimes have wet summers and dry winters. Tropical rainforests technically do not have dry or wet seasons, since their rainfall is equally distributed through the year. Some areas with pronounced rainy seasons will see a break in rainfall mid-season when the intertropical convergence zone or monsoon trough move poleward of their location during the middle of the warm season. When the wet season occurs during the warm season, or summer, rain falls mainly during the late afternoon and early evening hours. The wet season is a time when air quality improves, freshwater quality improves, and vegetation grows significantly.

Tropical cyclones, a source of very heavy rainfall, consist of large air masses several hundred miles across with low pressure at the centre and with winds blowing inward towards the centre in either a clockwise direction (southern hemisphere) or counter clockwise (northern hemisphere). Although cyclones can take an enormous toll in lives and personal property, they may be important factors in the precipitation regimes of places they impact, as they may bring much-needed precipitation to otherwise dry regions. Areas in their path can receive a year's worth of rainfall from a tropical cyclone passage.

Human Influence

The fine particulate matter produced by car exhaust and other human sources of pollution forms cloud condensation nuclei, leads to the production of clouds and increases the likelihood of rain. As commuters and commercial traffic cause pollution to build up over the course of the week, the likelihood of rain increases: it peaks by Saturday, after five days of weekday pollution has been built up. In heavily populated areas that are near the coast, such as the United States' Eastern Seaboard, the effect can be dramatic: there is a 22% higher chance of rain on Saturdays than on Mondays. The urban heat island effect warms cities 0.6 °C (1.1 °F) to 5.6 °C (10.1 °F) above surrounding suburbs and rural areas. This extra heat leads to greater upward motion, which can induce additional shower and thunderstorm activity. Rainfall rates downwind of cities are increased between 48% and 116%. Partly as a result of this warming, monthly rainfall is about 28% greater between 32 to 64 km (20 to 40 mi) downwind of cities, compared with upwind. Some cities induce a total precipitation increase of 51%.

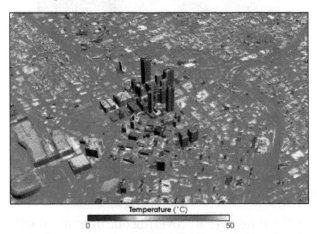

Image of Atlanta, Georgia showing temperature distribution, with blue showing cool temperatures, red warm, and hot areas appear white.

Mean surface temperature anomalies during the period 1999 to 2008 with respect to the average temperatures from 1940 to 1980

Increasing temperatures tend to increase evaporation which can lead to more precipitation. Precipitation generally increased over land north of 30°N from 1900 through 2005 but has declined over the tropics since the 1970s. Globally there has been no statistically significant overall trend in precipitation over the past century, although trends have varied widely by region and over time. Eastern portions of North and South America, northern Europe, and northern and central Asia have become wetter. The Sahel, the Mediterranean, southern Africa and parts of southern Asia have become drier. There has been an increase in the number of heavy precipitation events over many areas during the past century, as well as an increase since the 1970s in the prevalence of droughts—especially in the tropics and subtropics. Changes in precipitation and evaporation over the oceans are suggested by the decreased salinity of mid- and high-latitude waters (implying more precipitation), along with increased salinity in lower latitudes (implying less precipitation and/or more evaporation). Over the contiguous United States, total annual precipitation increased at an average rate of 6.1 percent since 1900, with the greatest increases within the East North Central climate region (11.6 percent per century) and the South (11.1 percent). Hawaii was the only region to show a decrease (−9.25 percent).

Analysis of 65 years of United States of America rainfall records show the lower 48 states have an increase in heavy downpours since 1950. The largest increases are in the Northeast and Midwest, which in the past decade, have seen 31 and 16 percent more heavy downpours compared to the 1950s. Rhode Island is the state with the largest increase 104%. McAllen, Texas is the city with the largest increase, 700%. Heavy downpour in the analysis are the days where total precipitation exceeded the top 1 percent of all rain and snow days during the years 1950-2014

The most successful attempts at influencing weather involve cloud seeding, which include techniques used to increase winter precipitation over mountains and suppress hail.

Characteristics

Patterns

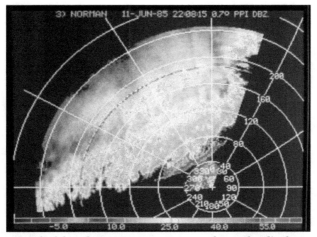

Band of thunderstorms seen on a weather radar display

Rainbands are cloud and precipitation areas which are significantly elongated. Rainbands can be stratiform or convective, and are generated by differences in temperature. When noted on weather radar imagery, this precipitation elongation is referred to as banded structure. Rainbands in advance of warm occluded fronts and warm fronts are associated with weak upward motion, and tend to be wide and stratiform in nature.

Rainbands spawned near and ahead of cold fronts can be squall lines which are able to produce tornadoes. Rainbands associated with cold fronts can be warped by mountain barriers perpendicular to the front's orientation due to the formation of a low-level barrier jet. Bands of thunderstorms can form with sea breeze and land breeze boundaries, if enough moisture is present. If sea breeze rainbands become active enough just ahead of a cold front, they can mask the location of the cold front itself.

Once a cyclone occludes, a trough of warm air aloft, or "trowal" for short, will be caused by strong southerly winds on its eastern periphery rotating aloft around its northeast, and ultimately northwestern, periphery (also known as the warm conveyor belt), forcing a surface trough to continue into the cold sector on a similar curve to the occluded front. The trowal creates the portion of an occluded cyclone known as its comma head, due to the comma-like shape of the mid-tropospheric cloudiness that accompanies the feature. It can also be the focus of locally heavy precipitation, with thunderstorms possible if the atmosphere along the trowal is unstable enough for convection. Banding within the comma head precipitation pattern of an extratropical cyclone can yield significant amounts of rain. Behind extratropical cyclones during fall and winter, rainbands can form downwind of relative warm bodies of water such as the Great Lakes. Downwind of islands, bands of showers and thunderstorms can develop due to low level wind convergence downwind of the island edges. Offshore California, this has been noted in the wake of cold fronts.

Rainbands within tropical cyclones are curved in orientation. Tropical cyclone rainbands contain showers and thunderstorms that, together with the eyewall and the eye, constitute a hurricane or tropical storm. The extent of rainbands around a tropical cyclone can help determine the cyclone's intensity.

Acidity

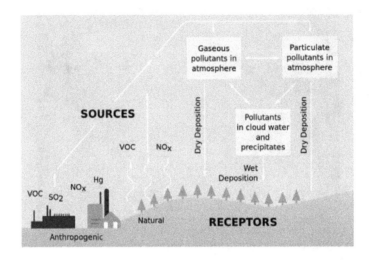

The phrase *acid rain* was first used by Scottish chemist Robert Augus Smith in 1852. The pH of rain varies, especially due to its origin. On America's East Coast, rain that is derived from the Atlantic Ocean typically has a pH of 5.0-5.6; rain that comes across the continental from the west has a pH of 3.8-4.8; and local thunderstorms can have a pH as low as 2.0. Rain becomes acidic primarily due to the presence of two strong acids, sulfuric acid (H_2SO_4) and nitric acid (HNO_3). Sulfuric acid is derived from natural sources such as volcanoes, and wetlands (sulfate reducing bacteria); and anthropogenic sources such as the combustion of fossil fuels, and mining where H_2S is present. Nitric acid is produced by natural sources such as lightning, soil bacteria, and natural fires; while also produced anthropogenically by the combustion of fossil fuels and from power plants. In the past 20 years the concentrations of nitric and sulfuric acid has decreased in presence of rainwater, which may be due to the significant increase in ammonium (most likely as ammonia from livestock production), which acts as a buffer in acid rain and raises the pH.

Köppen Climate Classification

The Köppen classification depends on average monthly values of temperature and precipitation. The most commonly used form of the Köppen classification has five primary types labeled A through E. Specifically, the primary types are A, tropical; B, dry; C, mild mid-latitude; D, cold mid-latitude; and E, polar. The five primary classifications can be further divided into secondary classifications such as rain forest, monsoon, tropical savanna, humid subtropical, humid continental, oceanic climate, Mediterranean climate, steppe, subarctic climate, tundra, polar ice cap, and desert.

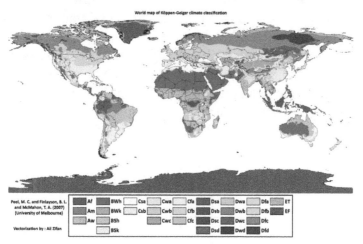

Updated Köppen-Geiger climate map

Rain forests are characterized by high rainfall, with definitions setting minimum normal annual rainfall between 1,750 and 2,000 mm (69 and 79 in). A tropical savanna is a grassland biome located in semi-arid to semi-humid climate regions of subtropical and tropical latitudes, with rainfall between 750 and 1,270 mm (30 and 50 in) a year. They are widespread on Africa, and are also found in India, the northern parts of South America, Malaysia, and Australia. The humid subtropical climate zone where winter rainfall is associated with large storms that the westerlies steer from west to east. Most summer rainfall occurs during thunderstorms and from occasional tropical cyclones. Humid subtropical climates lie on the east side continents, roughly between latitudes 20° and 40° degrees away from the equator.

An oceanic (or maritime) climate is typically found along the west coasts at the middle latitudes of all the world's continents, bordering cool oceans, as well as southeastern Australia, and is accompanied by plentiful precipitation year round. The Mediterranean climate regime resembles the climate of the lands in the Mediterranean Basin, parts of western North America, parts of Western and South Australia, in southwestern South Africa and in parts of central Chile. The climate is characterized by hot, dry summers and cool, wet winters. A steppe is a dry grassland. Subarctic climates are cold with continuous permafrost and little precipitation.

Measurement

Gauges

Rain is measured in units of length per unit time, typically in millimeters per hour, or in countries where imperial units are more common, inches per hour. The "length", or more accurately, "depth" being measured is the depth of rain water that would accumulate on a flat, horizontal and impermeable surface during a given amount of time, typically an hour. One millimeter of rainfall is the equivalent of one liter of water per square meter.

The standard way of measuring rainfall or snowfall is the standard rain gauge, which can be found in 100-mm (4-in) plastic and 200-mm (8-in) metal varieties. The inner cylinder is filled by 25 mm

(0.98 in) of rain, with overflow flowing into the outer cylinder. Plastic gauges have markings on the inner cylinder down to 0.25 mm (0.0098 in) resolution, while metal gauges require use of a stick designed with the appropriate 0.25 mm (0.0098 in) markings. After the inner cylinder is filled, the amount inside it is discarded, then filled with the remaining rainfall in the outer cylinder until all the fluid in the outer cylinder is gone, adding to the overall total until the outer cylinder is empty. Other types of gauges include the popular wedge gauge (the cheapest rain gauge and most fragile), the tipping bucket rain gauge, and the weighing rain gauge. For those looking to measure rainfall the most inexpensively, a can that is cylindrical with straight sides will act as a rain gauge if left out in the open, but its accuracy will depend on what ruler is used to measure the rain with. Any of the above rain gauges can be made at home, with enough know-how.

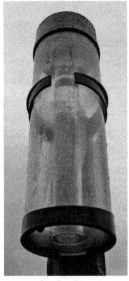

Standard rain gauge

When a precipitation measurement is made, various networks exist across the United States and elsewhere where rainfall measurements can be submitted through the Internet, such as CoCo-RAHS or GLOBE. If a network is not available in the area where one lives, the nearest local weather or met office will likely be interested in the measurement.

Remote Sensing

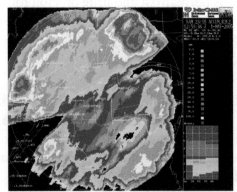

Twenty-four-hour rainfall accumulation on the Val d'Irène radar in Eastern Canada. Zones without data in the east and southwest are caused by beam blocking from mountains. (Source: Environment Canada)

One of the main uses of weather radar is to be able to assess the amount of precipitations fallen over large basins for hydrological purposes. For instance, river flood control, sewer management and dam construction are all areas where planners use rainfall accumulation data. Radar-derived rainfall estimates compliment surface station data which can be used for calibration. To produce radar accumulations, rain rates over a point are estimated by using the value of reflectivity data at individual grid points. A radar equation is then used, which is,

$$Z = A\,R^b$$

where Z represents the radar reflectivity, R represents the rainfall rate, and A and b are constants. Satellite derived rainfall estimates use passive microwave instruments aboard polar orbiting as well as geostationary weather satellites to indirectly measure rainfall rates. If one wants an accumulated rainfall over a time period, one has to add up all the accumulations from each grid box within the images during that time.

1988 rain in the U.S. The heaviest rain is seen in reds and yellows.

1993 rain in the U.S.

Intensity

Rainfall intensity is classified according to the rate of precipitation:

- Light rain — when the precipitation rate is < 2.5 mm (0.098 in) per hour

- Moderate rain — when the precipitation rate is between 2.5 mm (0.098 in) - 7.6 mm (0.30 in) or 10 mm (0.39 in) per hour

- Heavy rain — when the precipitation rate is > 7.6 mm (0.30 in) per hour, or between 10 mm (0.39 in) and 50 mm (2.0 in) per hour

- Violent rain — when the precipitation rate is > 50 mm (2.0 in) per hour

Euphemisms for a heavy or violent rain include gully washer, trash-mover and toad-strangler. The intensity can also be expressed by rainfall erosivity. R-factor

Return Period

The likelihood or probability of an event with a specified intensity and duration, is called the return period or frequency. The intensity of a storm can be predicted for any return period and storm duration, from charts based on historic data for the location. The term *1 in 10 year storm* describes a rainfall event which is rare and is only likely to occur once every 10 years, so it has a 10 percent likelihood any given year. The rainfall will be greater and the flooding will be worse than the worst storm expected in any single year. The term *1 in 100 year storm* describes a rainfall event which is extremely rare and which will occur with a likelihood of only once in a century, so has a 1 percent likelihood in any given year. The rainfall will be extreme and flooding to be worse than a 1 in 10 year event. As with all probability events, it is possible, though improbable, to have multiple "1 in 100 Year Storms" in a single year.

Forecasting

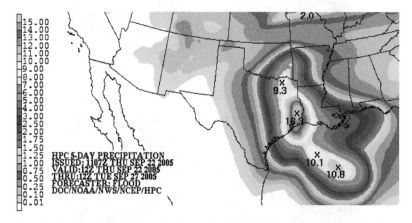

Example of a five-day rainfall forecast from the Hydrometeorological Prediction Center

The Quantitative Precipitation Forecast (abbreviated QPF) is the expected amount of liquid precipitation accumulated over a specified time period over a specified area. A QPF will be specified when a measurable precipitation type reaching a minimum threshold is forecast for any hour during a QPF valid period. Precipitation forecasts tend to be bound by synoptic hours such as 0000, 0600, 1200 and 1800 GMT. Terrain is considered in QPFs by use of topography or based upon climatological precipitation patterns from observations with fine detail. Starting in the mid to late 1990s, QPFs were used within hydrologic forecast models to simulate impact to rivers throughout the United States. Forecast models show significant sensitivity to humidity levels within the planetary boundary layer, or in the lowest levels of the atmosphere, which decreases with height. QPF can be generated on a quantitative, forecasting amounts, or a qualitative, forecasting the probability of a specific amount, basis. Radar imagery forecasting techniques show higher skill than model forecasts within 6 to 7 hours of the time of the radar image. The forecasts can be verified through use of rain gauge measurements, weather radar estimates, or a combination of both. Various skill scores can be determined to measure the value of the rainfall forecast.

Impact

Effect on Agriculture

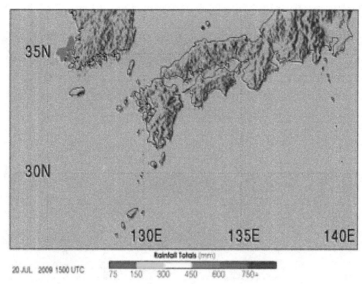

Rainfall estimates for southern Japan and the surrounding region from July 20–27, 2009.

Precipitation, especially rain, has a dramatic effect on agriculture. All plants need at least some water to survive, therefore rain (being the most effective means of watering) is important to agriculture. While a regular rain pattern is usually vital to healthy plants, too much or too little rainfall can be harmful, even devastating to crops. Drought can kill crops and increase erosion, while overly wet weather can cause harmful fungus growth. Plants need varying amounts of rainfall to survive. For example, certain cacti require small amounts of water, while tropical plants may need up to hundreds of inches of rain per year to survive.

In areas with wet and dry seasons, soil nutrients diminish and erosion increases during the wet season. Animals have adaptation and survival strategies for the wetter regime. The previous dry season leads to food shortages into the wet season, as the crops have yet to mature. Developing countries have noted that their populations show seasonal weight fluctuations due to food shortages seen before the first harvest, which occurs late in the wet season. Rain may be harvested through the use of rainwater tanks; treated to potable use or for non-potable use indoors or for irrigation. Excessive rain during short periods of time can cause flash floods.

In Culture

Cultural attitudes towards rain differ across the world. In temperate climates, people tend to be more stressed when the weather is unstable or cloudy, with its impact greater on men than women. Rain can also bring joy, as some consider it to be soothing or enjoy the aesthetic appeal of it. In dry places, such as India, or during periods of drought, rain lifts people's moods. In Botswana, the Setswana word for rain, *pula*, is used as the name of the national currency, in recognition of the economic importance of rain in this desert country. Several cultures have developed means of dealing with rain and have developed numerous protection devices such as umbrellas and raincoats, and diversion devices such as gutters and storm drains that lead rains to sewers. Many people find the scent during and immediately after rain pleasant or distinctive. The source of this scent is petrichor, an oil produced by plants, then absorbed by rocks and soil, and later released into the air during rainfall.

Global Climatology

Approximately 505,000 km³ (121,000 cu mi) of water falls as precipitation each year across the globe with 398,000 km³ (95,000 cu mi) of it over the oceans. Given the Earth's surface area, that means the globally averaged annual precipitation is 990 mm (39 in). Deserts are defined as areas with an average annual precipitation of less than 250 mm (10 in) per year, or as areas where more water is lost by evapotranspiration than falls as precipitation.

Deserts

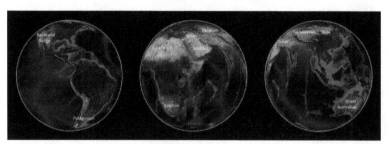

Largest deserts

The northern half of Africa is occupied by the world's most extensive hot, dry region, the Sahara Desert. Some deserts are also occupying much of southern Africa : the Namib and the Kalahari. Across Asia, a large annual rainfall minimum, composed primarily of deserts, stretches from the Gobi Desert in Mongolia west-southwest through western Pakistan (Balochistan) and Iran into the Arabian Desert in Saudi Arabia. Most of Australia is semi-arid or desert, making it the world's driest inhabited continent. In South America, the Andes mountain range blocks Pacific moisture that arrives in that continent, resulting in a desertlike climate just downwind across western Argentina. The drier areas of the United States are regions where the Sonoran Desert overspreads the Desert Southwest, the Great Basin and central Wyoming.

Isolated towering vertical desert shower

Polar Desert

Since rain only falls as liquid, in frozen temperatures, rain cannot fall. As a result, very cold climates see very little rainfall and are often known as polar deserts. A common biome in this area is the tundra which has a short summer thaw and a long frozen winter. Ice caps see no rain at all, making Antarctica the world's driest continent.

Rainforests

Rainforests are areas of the world with very high rainfall. Both tropical and temperate rainforests exist. Tropical rainforests occupy a large band of the planet mostly along the equator. Most temperate rainforests are located on mountainous west coasts between 45 and 55 degrees latitude, but they are often found in other areas.

Around 40-75% of all biotic life is found in rainforests. Rainforests are also responsible for 28% of the world's oxygen turnover.

Monsoons

The equatorial region near the Intertropical Convergence Zone (ITCZ), or monsoon trough, is the wettest portion of the world's continents. Annually, the rain belt within the tropics marches northward by August, then moves back southward into the Southern Hemisphere by February and March. Within Asia, rainfall is favored across its southern portion from India east and northeast across the Philippines and southern China into Japan due to the monsoon advecting moisture primarily from the Indian Ocean into the region. The monsoon trough can reach as far north as the 40th parallel in East Asia during August before moving southward thereafter. Its poleward progression is accelerated by the onset of the summer monsoon which is characterized by the development of lower air pressure (a thermal low) over the warmest part of Asia. Similar, but weaker, monsoon circulations are present over North America and Australia. During the summer, the Southwest monsoon combined with Gulf of California and Gulf of Mexico moisture moving around the subtropical ridge in the Atlantic ocean bring the promise of afternoon and evening thunderstorms to the southern tier of the United States as well as the Great Plains. The eastern half of the contiguous United States east of the 98th meridian, the mountains of the Pacific Northwest, and the Sierra Nevada range are the wetter portions of the nation, with average rainfall exceeding 760 mm (30 in) per year. Tropical cyclones enhance precipitation across southern sections of the United States, as well as Puerto Rico, the United States Virgin Islands, the Northern Mariana Islands, Guam, and American Samoa.

Impact of The Westerlies

Westerly flow from the mild north Atlantic leads to wetness across western Europe, in particular Ireland and the United Kingdom, where the western coasts can receive between 1,000 mm (39 in), at sea-level and 2,500 mm (98 in), on the mountains of rain per year. Bergen, Norway is one of the more famous European rain-cities with its yearly precipitation of 2,250 mm (89 in) on average. During the fall, winter, and spring, Pacific storm systems bring most of Hawaii and the western United States much of their precipitation. Over the top of the ridge, the jet stream brings a summer precipitation maximum to the Great Lakes. Large thunderstorm areas known as mesoscale convective complexes move through the Plains, Midwest, and Great Lakes during the warm season, contributing up to 10% of the annual precipitation to the region.

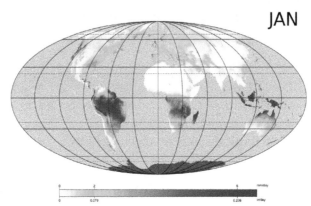

JAN

Long-term mean precipitation by month

The El Niño-Southern Oscillation affects the precipitation distribution, by altering rainfall patterns across the western United States, Midwest, the Southeast, and throughout the tropics. There is also evidence that global warming is leading to increased precipitation to the eastern portions of North America, while droughts are becoming more frequent in the tropics and subtropics.

Wettest Known Locations

Cherrapunji, situated on the southern slopes of the Eastern Himalaya in Shillong, India is the confirmed wettest place on Earth, with an average annual rainfall of 11,430 mm (450 in). The highest recorded rainfall in a single year was 22,987 mm (905.0 in) in 1861. The 38-year average at nearby Mawsynram, Meghalaya, India is 11,873 mm (467.4 in). The wettest spot in Australia is Mount Bellenden Ker in the north-east of the country which records an average of 8,000 mm (310 in) per year, with over 12,200 mm (480.3 in) of rain recorded during 2000. Mount Wai'ale' ale on the island of Kaua'i in the Hawaiian Islands averages more than 12,000 mm (460 in) of rain per year over the last 32 years, with a record 17,340 mm (683 in) in 1982.Its summit is considered one of the rainiest spots on earth. It has been promoted in tourist literature for many years as the wettest spot in the world. Lloró, a town situated in Chocó, Colombia, is probably the place with the largest rainfall in the world, averaging 13,300 mm (523.6 in) per year. The Department of Chocó is extraordinarily humid. Tutunendaó, a small town situated in the same department, is one of the wettest estimated places on Earth, averaging 11,394 mm (448.6 in) per year; in 1974 the town received 26,303 mm (86 ft 3.6 in),the largest annual rainfall measured in Colombia. Unlike Cherrapunji, which receives most of its rainfall between April and September, Tutunendaó receives rain almost uniformly distributed throughout the year. Quibdó, the capital of Chocó, receives the most rain in the world among cities with over 100,000 inhabitants: 9,000 mm (354 in) per year. Storms in Chocó can drop 500 mm (20 in) of rainfall in a day. This amount is more than what falls in many cities in a year's time.

Continent	Highest average		Place	Elevation		Years of record
	in	mm		ft	m	
South America	523.6	13,299	Lloró, Colombia (estimated)	520	158[c]	29
Asia	467.4	11,872	Mawsynram, India	4,597	1,401	39

Oceania	460.0	11,684	Mount Wai'ale'ale, Kauai, Hawaii (USA)	5,148	1,569	30
Africa	405.0	10,287	Debundscha, Cameroon	30	9.1	32
South America	354.0	8,992	Quibdo, Colombia	120	36.6	16
Australia	340.0	8,636	Mount Bellenden Ker, Queensland	5,102	1,555	9
North America	256.0	6,502	Henderson Lake, British Columbia	12	3.66	14
Europe	183.0	4,648	Crkvice, Montenegro	3,337	1,017	22
Source (without conversions): *Global Measured Extremes of Temperature and Precipitation*, National Climatic Data Center. August 9, 2004.						

	Continent	Place	Highest rainfall	
			in	mm
Highest average annual rainfall	Asia	Mawsynram, India	467.4	11,870
Highest in one year	Asia	Cherrapunji, India	1,042	26,470
Highest in one calendar month	Asia	Cherrapunji, India	366	9,296
Highest in 24 hours	Indian Ocean	Foc Foc, La Reunion Island	71.8	1,820
Highest in 12 hours	Indian Ocean	Foc Foc, La Reunion Island	45.0	1,140
Highest in one minute	North America	Unionville, Maryland, USA	1.23	31.2

Outside of Earth

On Titan, Saturn's largest natural satellite, infrequent methane rain is thought to carve the moon's numerous surface channels. On Venus, sulfuric acid virga evaporates 25 km (16 mi) from the surface. There is likely to be rain of various compositions in the upper atmospheres of the gas giants, as well as precipitation of liquid neon in the deep atmospheres. Extrasolar planet OGLE-TR-56b in the constellation Sagittarius is hypothesized to have iron rain.

Cyclone

In meteorology, a cyclone is a large scale air mass that rotates around a strong center of low atmospheric pressure. They are usually characterized by inward spiraling winds that rotate counterclockwise in the Northern Hemisphere and clockwise in the Southern Hemisphere. All large-scale cyclones are centered on low-pressure areas. The largest low-pressure systems are polar vortices and extratropical cyclones of the largest scale (synoptic scale). According to the National Hurricane Center glossary, warm-core cyclones such as tropical cyclones and subtropical cyclones also lie within the synoptic scale. Mesocyclones, tornadoes and dust devils lie within the smaller mesoscale. Upper level cyclones can exist without the presence of a surface low, and can pinch off from the base of the Tropical Upper Tropospheric Trough during the summer months in the Northern Hemisphere. Cyclones have also been seen on extraterrestrial planets, such as Mars and Neptune. Cyclogenesis describes the process of cyclone formation and intensification. Extratropical cyclones begin as waves in large regions of enhanced mid-latitude temperature contrasts called baroclinic zones. These zones contract to form weather fronts as the cyclonic circulation closes and intensifies. Later in their life cycle, extratropical cyclones occlude as the cold air mass undercuts the warmer air and become cold core systems. A cyclone's track is guided over the course of its 2 to 6 day life cycle by the steering flow of the subtropical jet stream.

An extratropical cyclone near Iceland on September 4, 2003

Weather fronts separate two masses of air of different densities and are associated with the most prominent meteorological phenomena. Air masses separated by a front may differ in temperature or humidity. Strong cold fronts typically feature narrow bands of thunderstorms and severe weather, and may on occasion be preceded by squall lines or dry lines. They form west of the circulation center and generally move from west to east. Warm fronts form east of the cyclone center and are usually preceded by stratiform precipitation and fog. They move poleward ahead of the cyclone path. Occluded fronts form late in the cyclone life cycle near the center of the cyclone and often wrap around the storm center.

Tropical cyclogenesis describes the process of development of tropical cyclones. Tropical cyclones form due to latent heat driven by significant thunderstorm activity, and are warm core. Cyclones can transition between extratropical, subtropical, and tropical phases. Mesocyclones form as warm core cyclones over land, and can lead to tornado formation. Waterspouts can also form from mesocyclones, but more often develop from environments of high instability and low vertical wind shear. In the Atlantic and the northeastern Pacific oceans, a tropical cyclone is generally referred to as a hurricane (from the name of the ancient Central American deity of wind, Huracan), in the Indian and south Pacific oceans it is called a cyclone, and in the northwestern Pacific it is called a typhoon.

There are a number of structural characteristics common to all cyclones. A cyclone is a low-pressure area. A cyclone's center (often known in a mature tropical cyclone as the eye), is the area of lowest atmospheric pressure in the region. Near the center, the pressure gradient force (from the pressure in the center of the cyclone compared to the pressure outside the cyclone) and the force from the Coriolis effect must be in an approximate balance, or the cyclone would collapse on itself as a result of the difference in pressure.

Because of the Coriolis effect, the wind flow around a large cyclone is counterclockwise in the

Northern Hemisphere and clockwise in the Southern Hemisphere. In the Northern Hemisphere, the fastest winds relative to the surface of the Earth therefore occur on the eastern side of a northward-moving cyclone and on the northern side of a westward-moving one; the opposite occurs in the Southern Hemisphere. In contrast to low pressure systems, the wind flow around high pressure systems are clockwise (anticyclonic) in the northern hemisphere, and counterclockwise in the southern hemisphere.

Formation

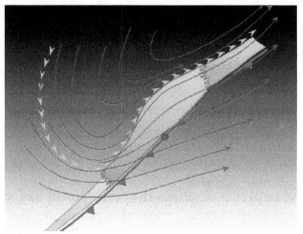

The initial extratropical low-pressure area forms at the location of the red dot on the image. It is usually perpendicular (at a right angle to) the leaf-like cloud formation seen on satellite during the early stage of cyclogenesis. The location of the axis of the upper level jet stream is in light blue.

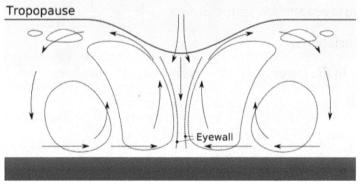

Tropical cyclones form when the energy released by the condensation of moisture in rising air causes a positive feedback loop over warm ocean waters.

Cyclogenesis is the development or strengthening of cyclonic circulation in the atmosphere. Arctic Climatology and Meteorology. Cyclogenesis is an umbrella term for several different processes, all of which result in the development of some sort of cyclone. It can occur at various scales, from the microscale to the synoptic scale.

Extratropical cyclones begin as waves along weather fronts before occluding later in their life cycle as cold-core systems. However, some intense extratropical cyclones can become warm-core systems when a warm seclusion occurs.

Tropical cyclones form as a result of significant convective activity, and are warm core. Mesocyclones form as warm core cyclones over land, and can lead to tornado formation. Waterspouts can also form from mesocyclones, but more often develop from environments of high instability and low vertical wind shear. Cyclolysis is the opposite of cyclogenesis, and is the high-pressure system equivalent which deals with the formation of high-pressure areas—Anticyclogenesis.

A surface low can form in a variety of ways. Topography can create a surface low. Mesoscale convective systems can spawn surface lows that are initially warm core. The disturbance can grow into a wave-like formation along the front and the low will be positioned at the crest. Around the low, the flow will become cyclonic. This rotational flow will move polar air will equatorward on the west side of the low, while warm air will move poleward on the east side. A cold front will appear on the west side, while a warm front will form on the east side. Usually the cold front will move at a quicker pace than the warm front and will "catch up" with it due to the slow erosion of higher density air mass located out ahead of the cyclone. In addition, the higher density air mass sweeping in behind the cyclone strengthens the higher pressure, denser cold air mass. The cold front over takes the warm front, and reduces the length of the warm front. At this point an occluded front forms where the warm air mass is pushed upwards into a trough of warm air aloft, which is also known as a trowal.

Tropical cyclogenesis is the term that describes the development and strengthening of a tropical cyclone. The mechanisms by which tropical cyclogenesis occurs are distinctly different from those that produce mid-latitude cyclones. Tropical cyclogenesis, the development of a warm-core cyclone, begins with significant convection in a favorable atmospheric environment. There are six main requirements for tropical cyclogenesis:

1. sufficiently warm sea surface temperatures,
2. atmospheric instability,
3. high humidity in the lower to middle levels of the troposphere
4. enough Coriolis force to develop a low-pressure center
5. a preexisting low-level focus or disturbance
6. low vertical wind shear.

An average of 86 tropical cyclones of tropical storm intensity form annually worldwide, with 47 reaching hurricane/typhoon strength, and 20 becoming intense tropical cyclones.

Synoptic Scale

The following types of cyclones are identifiable in synoptic charts.

Surface-based types

There are three main types surface-based cyclones: Extratropical cyclones, Subtropical cyclones and Tropical cyclones

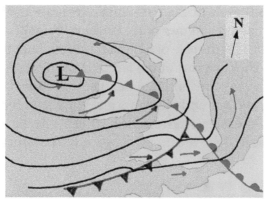

A fictitious synoptic chart of an extratropical cyclone affecting the UK and Ireland. The blue arrows between isobars indicate the direction of the wind, while the "L" symbol denotes the centre of the "low". Note the occluded, cold and warm frontal boundaries.

Extratropical Cyclone

An extratropical cyclone is a synoptic scale low-pressure weather system that does not have tropical characteristics, as it is connected with fronts and horizontal gradients (rather than vertical) in temperature and dew point otherwise known as "baroclinic zones".

"Extratropical" is applied to cyclones outside the tropics, in the middle latitudes. These systems may also be described as "mid-latitude cyclones" due to their area of formation, or "post-tropical cyclones" when a tropical cyclone has moved (extratropical transition) beyond the tropics. They are often described as "depressions" or "lows" by weather forecasters and the general public. These are the everyday phenomena which along with anti-cyclones, drive the weather over much of the Earth.

Although extratropical cyclones are almost always classified as baroclinic since they form along zones of temperature and dewpoint gradient within the westerlies, they can sometimes become barotropic late in their life cycle when the temperature distribution around the cyclone becomes fairly uniform with radius. An extratropical cyclone can transform into a subtropical storm, and from there into a tropical cyclone, if it dwells over warm waters sufficient to warm its core, and as a result develops central convection. A particularly intense type of extratropical cyclone that strikes during winter is known colloquially as a *nor'easter*.

Polar Low

A polar low is a small-scale, short-lived atmospheric low-pressure system (depression) that is found over the ocean areas poleward of the main polar front in both the Northern and Southern Hemispheres. Polar lows are cold-core so they can be considered as a subset of extratropical cyclones. Polar lows were first identified on the meteorological satellite imagery that became available in the 1960s, which revealed many small-scale cloud vortices at high latitudes. The most active polar lows are found over certain ice-free maritime areas in or near the Arctic during the winter, such as the Norwegian Sea, Barents Sea, Labrador Sea and Gulf of Alaska. Polar lows dissipate rapidly when they make landfall. Antarctic systems tend to be weaker than their northern counterparts since the air-sea temperature differences around the continent are generally smaller. However, vigorous polar lows can be found over the Southern Ocean. During winter, when cold-core lows with temperatures in the mid-levels of the troposphere reach −45 °C (−49 °F) move over open waters, deep convection forms which allows polar low development to become possible. The systems usually have a horizontal length scale of less than 1,000 kilometres (620 mi) and exist for no more than a couple of days. They are part of the larger class of mesoscale weather systems. Polar lows can be difficult to detect using conventional weather reports and are a hazard to high-latitude operations, such as shipping and gas and oil platforms. Polar lows have been referred to by many other terms, such as polar mesoscale vortex, Arctic hurricane, Arctic low, and cold air depression. Today the term is usually reserved for the more vigorous systems that have near-surface winds of at least 17 m/s.

Polar low over the Barents Sea on February 27, 1987

Subtropical

Subtropical Storm Andrea in 2007

A subtropical cyclone is a weather system that has some characteristics of a tropical cyclone and some characteristics of an extratropical cyclone. They can form between the equator and the 50th parallel. As early as the 1950s, meteorologists were unclear whether they should be characterized as tropical cyclones or extratropical cyclones, and used terms such as quasi-tropical and semi-tropical to describe the cyclone hybrids. By 1972, the National Hurricane Center officially recognized this cyclone category. Subtropical cyclones began to receive names off the official tropical cyclone list in the Atlantic Basin in 2002. They have broad wind patterns with maximum sustained winds located farther from the center than typical tropical cyclones, and exist in areas of weak to moderate temperature gradient.

Since they form from extratropical cyclones which have colder temperatures aloft than normally found in the tropics, the sea surface temperatures required is around 23 degrees Celsius (73 °F) for their formation, which is three degrees Celsius (5 °F) lower than for tropical cyclones. This means that subtropical cyclones are more likely to form outside the traditional bounds of the hurricane season. Although subtropical storms rarely have hurricane-force winds, they may become tropical in nature as their cores warm.

Tropical

2015 Atlantic hurricane season summary map

A tropical cyclone is a storm system characterized by a low-pressure center and numerous thunderstorms that produce strong winds and flooding rain. A tropical cyclone feeds on heat released when moist air rises, resulting in condensation of water vapour contained in the moist air. They are fueled by a different heat mechanism than other cyclonic windstorms such as nor'easters, European windstorms, and polar lows, leading to their classification as "warm core" storm systems.

Cyclone Catarina, a rare South Atlantic tropical cyclone viewed from the International Space Station on March 26, 2004

The term "tropical" refers to both the geographic origin of these systems, which form almost exclusively in tropical regions of the globe, and their dependence on Maritime Tropical air masses for their formation. The term "cyclone" refers to the storms' cyclonic nature, with counterclockwise rotation in the Northern Hemisphere and clockwise rotation in the Southern Hemisphere. Depending on their location and strength, tropical cyclones are referred to by other names, such as hurricane, typhoon, tropical storm, cyclonic storm, tropical depression, or simply as a cyclone.

While tropical cyclones can produce extremely powerful winds and torrential rain, they are also able to produce high waves and a damaging storm surge. Their winds increase the wave size, and in so doing they draw more heat and moisture into their system, thereby increasing their strength. They develop over large bodies of warm water, and hence lose their strength if they move over land. This is the reason coastal regions can receive significant damage from a tropical cyclone, while inland regions are relatively safe from strong winds. Heavy rains, however, can produce significant flooding inland. Storm surges are rises in sea level caused by the reduced pressure of the core that in effect "sucks" the water upward and from winds that in effect "pile" the water up. Storm surges can produce extensive coastal flooding up to 40 kilometres (25 mi) from the coastline. Although their effects on human populations can be devastating, tropical cyclones can also relieve drought conditions. They also carry heat and energy away from the tropics and transport it toward temperate latitudes, which makes them an important part of the global atmospheric circulation mechanism. As a result, tropical cyclones help to maintain equilibrium in the Earth's troposphere.

Many tropical cyclones develop when the atmospheric conditions around a weak disturbance in the atmosphere are favorable. Others form when other types of cyclones acquire tropical characteristics. Tropical systems are then moved by steering winds in the troposphere; if the conditions remain favorable, the tropical disturbance intensifies, and can even develop an eye. On the other end of the spectrum, if the conditions around the system deteriorate or the tropical cyclone makes landfall, the system will weaken and eventually dissipate. A tropical cyclone can become extratropical as it moves toward higher latitudes if its energy source changes from heat released by condensation to differences in temperature between air masses. A tropical cyclone is usually not considered to become subtropical during its extratropical transition.

Upper level types

Polar Cyclone

A polar, sub-polar, or Arctic cyclone (also known as a polar vortex) is a vast area of low pressure which strengthens in the winter and weakens in the summer. A polar cyclone is a low-pressure weather system, usually spanning 1,000 kilometres (620 mi) to 2,000 kilometres (1,200 mi), in which the air circulates in a counterclockwise direction in the northern hemisphere, and a clockwise direction in the southern hemisphere. The Coriolis acceleration acting on the air masses moving poleward at high altitude, causes a counterclockwise circulation at high altitude. The poleward movement of air originates from the air circulation of the Polar cell. The polar low is not driven by convection as are tropical cyclones, nor the cold and warm air mass interactions as are extratropical cyclones, but is an artifact of the global air movement of the Polar cell. The base of the polar low is in the mid to upper troposphere. In the Northern Hemisphere, the polar cyclone has two centers on average. One center lies near Baffin Island and the other over northeast Siberia. In the southern hemisphere, it tends to be located near the edge of the Ross ice shelf near 160 west longitude. When the polar vortex is strong, its effect can be felt at the surface as a westerly wind (toward the east). When the polar cyclone is weak, significant cold outbreaks occur.

TUTT Cell

Under specific circumstances, upper level cold lows can break off from the base of the Tropical Upper Tropospheric Trough (TUTT), which is located mid-ocean in the Northern Hemisphere during the summer months. These upper tropospheric cyclonic vortices, also known as TUTT cells or TUTT lows, usually move slowly from east-northeast to west-southwest, and their bases generally do not extend below 20,000 feet in altitude. A weak inverted surface trough within the trade wind is generally found underneath them, and they may also be associated with broad areas of high-level clouds. Downward development results in an increase of cumulus clouds and the appearance of a surface vortex. In rare cases, they become warm-core tropical cyclones. Upper cyclones and the upper troughs which trail tropical cyclones can cause additional outflow channels and aid in their intensification. Developing tropical disturbances can help create or deepen upper troughs or upper lows in their wake due to the outflow jet emanating from the developing tropical disturbance/cyclone.

Mesoscale

The following types of cyclones are not identifiable in synoptic charts.

Mesocyclone

A mesocyclone is a vortex of air, 2.0 kilometres (1.2 mi) to 10 kilometres (6.2 mi) in diameter (the mesoscale of meteorology), within a convective storm. Air rises and rotates around a vertical axis, usually in the same direction as low-pressure systems in both northern and southern hemisphere. They are most often cyclonic, that is, associated with a localized low-pressure region within a supercell. Such storms can feature strong surface winds and severe hail. Mesocyclones often occur together with updrafts in supercells, where tornadoes may form. About 1700 mesocyclones form annually across the United States, but only half produce tornadoes.

Tornado

A tornado is a violently rotating column of air that is in contact with both the surface of the earth and a cumulonimbus cloud or, in rare cases, the base of a cumulus cloud. Also referred to as twisters, a colloquial term in America, or cyclones, although the word cyclone is used in meteorology, in a wider sense, to name any closed low-pressure circulation.

Dust Devil

A dust devil is a strong, well-formed, and relatively long-lived whirlwind, ranging from small (half a metre wide and a few metres tall) to large (more than 10 metres wide and more than 1000 metres tall). The primary vertical motion is upward. Dust devils are usually harmless, but can on rare occasions grow large enough to pose a threat to both people and property.

Waterspout

A waterspout is a columnar vortex forming over water that is, in its most common form, a non-supercell tornado over water that is connected to a cumuliform cloud. While it is often weaker than most of its land counterparts, stronger versions spawned by mesocyclones do occur.

Steam Devil

A gentle vortex over calm water or wet land made visible by rising water vapour.

Fire whirl

A fire whirl – also colloquially known as a fire devil, fire tornado, firenado, or fire twister – is a whirlwind induced by a fire and often made up of flame or ash.

Climate Change

Scientists warn that climate change could increase the intensity of typhoons as climate change projections show that the difference in temperature between the ocean – the heat source for cyclones – and the storm tops – the cold parts of cyclones – are likely to increase. Climate change is predicted to increase the frequency of high-intensity storms in selected ocean basins. While the effect changing climate is having on tropical storms remains largely unresolved scientists and president of Vanuatu Baldwin Lonsdale say the devastation caused by Pam, was aggravated by climate change.

Other Planets

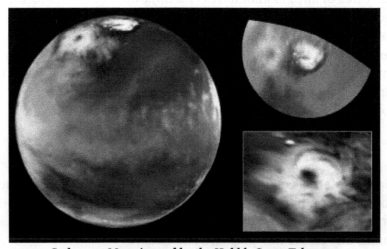

Cyclone on Mars, imaged by the Hubble Space Telescope

Cyclones are not unique to Earth. Cyclonic storms are common on Jovian planets, such as the Small Dark Spot on Neptune. It is about one third the diameter of the Great Dark Spot and received the nickname "Wizard's Eye" because it looks like an eye. This appearance is caused by a white cloud in the middle of the Wizard's Eye. Mars has also exhibited cyclonic storms. Jovian storms like the Great Red Spot are usually mistakenly named as giant hurricanes or cyclonic storms. However, this is inaccurate, as the Great Red Spot is, in fact, the inverse phenomenon, an anticyclone.

Cyclogenesis

Cyclogenesis is the development or strengthening of cyclonic circulation in the atmosphere (a low-pressure area). Cyclogenesis is an umbrella term for at least three different processes, all of which result in the development of some sort of cyclone, and at any size from the microscale to the synoptic scale.

- Tropical cyclones form due to latent heat driven by significant thunderstorm activity, and are warm core.

- Extratropical cyclones form as waves along weather fronts before occluding later in their life cycle as cold core cyclones.

- Mesocyclones form as warm core cyclones over land, and can lead to tornado formation. Waterspouts can also form from mesocyclones, but more often develop from environments of high instability and low vertical wind shear.

The process in which an extratropical cyclone undergoes a rapid drop in atmospheric pressure (24 millibars or more) in a 24-hour period is referred to as explosive cyclogenesis, and is usually present during the formation of a nor'easter. The anticyclonic equivalent, the process of formation of high pressure areas, is anticyclogenesis. The opposite of cyclogenesis is cyclolysis.

Meteorological Scales

There are four main scales, or sizes of systems, dealt with in meteorology: the macroscale, the synoptic scale, the mesoscale, and the microscale. The macroscale deals with systems with global size, such as the Madden–Julian oscillation. Synoptic scale systems cover a portion of a continent, such as extratropical cyclones, with dimensions of 1,000-2,500 km (620-1,550 mi) across. The mesoscale is the next smaller scale, and often is divided into two ranges: meso-alpha phenomena range from 200-2,000 km (125-1,243 mi) across (the realm of the tropical cyclone), while meso-beta phenomena range from 20–200 km (12-125 mi) across (the scale of the mesocyclone). The microscale is the smallest of the meteorological scales, with a size under two kilometers (1.2 mi) (the scale of tornadoes and waterspouts). These horizontal dimensions are not rigid divisions but instead reflect typical sizes of phenomena having certain dynamic characteristics. For example, a system does not necessarily transition from meso-alpha to synoptic scale when its horizontal extent grows from 2,000 to 2,001 km (1,243 mi).

Extratropical Cyclones

The initial frontal wave (or low pressure area) forms at the location of the red dot on the image. It is usually perpendicular (at a right angle) to the leaf-like cloud formation (baroclinic leaf) seen on satellite during the early stage of cyclogenesis. The location of the axis of the upper level jet stream is in light blue.

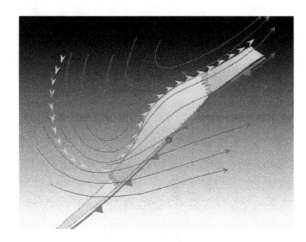

Norwegian Cyclone Model

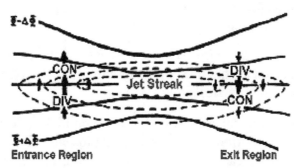

An upper level jet streak. DIV areas are regions of divergence aloft, which will lead to surface convergence and aid cyclogenesis.

The Norwegian Cyclone Model is an idealized formation model of cold-core cyclonic storms developed by Norwegian meteorologists during the First World War. The main concept behind this model, relating to cyclogenesis, is that cyclones progress through a predictable evolution as they move up a frontal boundary, with the most mature cyclone near the northeast end of the front and the least mature near the tail end of the front.

Precursors for Development

A preexisting frontal boundary, as defined in surface weather analysis, is required for the development of a mid-latitude cyclone. The cyclonic flow begins around a disturbed section of the stationary front due to an upper level disturbance, such as a short wave or an upper-level trough, near a favorable quadrant of the upper level jet. However, enhanced along-frontal stretching rates in the lower troposphere can suppress the growth of extratropical cyclones.

Vertical Motion Affecting Development

Cyclogenesis can only occur when temperature decreases polewards (to the north, in the northern hemisphere), and pressure perturbation lines tilt westward with height. Cyclogenesis is most likely to occur in regions of cyclonic vorticity advection, downstream of a strong westerly jet. The combination of vorticity advection and thermal advection created by the temperature gradient and a low pressure center cause upward motion around the low. If the temperature gradient is strong enough, temperature advection will increase, driving more vertical motion. This increases the overall strength of the system. Shearwise updrafts are the most important factor in determining cyclonic growth and strength.

Modes of Development

The surface low could have a variety of causes for forming. Topography can force a surface low when dense low-level high pressure system ridges in east of a north-south mountain barrier. Mesoscale convective systems can spawn surface lows which are initially warm core. The disturbance can grow into a wave-like formation along the front and the low will be positioned at the crest. Around the low, flow will become cyclonic, by definition. This rotational flow will push polar air equatorward west of the low via its trailing cold front, and warmer air will push poleward low via the warm front. Usually the cold front will move at a quicker pace than the warm front and "catch up" with it due to the slow erosion of higher density airmass located out ahead of the cyclone and the higher density airmass sweeping in behind the cyclone, usually resulting in a narrowing warm sector. At this point an occluded front forms where the warm air mass is pushed upwards into a trough of warm air aloft, which is also known as a trowal (a trough of warm air aloft). All developing low pressure areas share one important aspect, that of upward vertical motion within the troposphere. Such upward motions decrease the mass of local atmospheric columns of air, which lower surface pressure.

Maturity

Maturity is after the time of occlusion when the storm has completed strengthening and the cyclonic flow is at its most intense. Thereafter, the strength of the storm diminishes as the cyclone couples with the upper level trough or upper level low, becoming increasingly cold core. The spin-down of cyclones, also known as cyclolysis, can be understood from an energetics perspective. As occlusion occurs and the warm air mass is pushed upwards over a cold air airmass, the atmosphere becomes increasingly stable and the centre of gravity of the system lowers. As the occlusion process extends further down the warm front and away from the central low, more and more of the available potential energy of the system is exhausted. This potential energy sink creates a kinetic energy source which injects a final burst of energy into the storm's motions. After this process occurs, the growth period of the cyclone, or cyclogenesis, ends, and the low begins to spin down (fill) as more air is converging into the bottom of the cyclone than is being removed out the top since upper-level divergence has decreased.

Occasionally, cyclogenesis will re-occur with occluded cyclones. When this happens a new low center will form on the triple-point (the point where the cold front, warm front, and occluded front meet). During triple-point cyclogenesis, the occluded parent low will fill as the secondary low deepens into the main weathermaker.

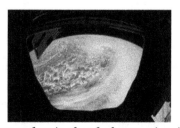

This image displays a dense cloud pattern and arcing band of convection, indicating a young, developing cyclone.

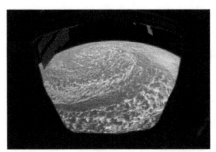

The diffuse cloud pattern in this image indicates an old, dissipating low pressure system.

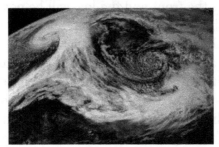

This image illustrates the relative positions of two storm systems over the North-eastern Pacific.

Tropical Cyclones

Tropical cyclones exist within a mesoscale alpha domain. As opposed to mid-latitude cyclogenesis, tropical cyclogenesis is driven by strong convection organised into a central core with no baroclinic zones, or fronts, extending through their center. Although the formation of tropical cyclones is the topic of extensive ongoing research and is still not fully understood, there are six main requirements for tropical cyclogenesis: sea surface temperatures that are warm enough, atmospheric instability, high humidity in lower to middle levels of the troposphere, enough Coriolis force to develop a low pressure center, a pre-existing low level focus or disturbance, and low vertical wind shear. These warm core cyclones tend to form over the oceans between 10 and 30 degrees of the equator.

Mesocyclones

Mesocyclones range in size from mesoscale beta to microscale. The term mesocyclone is usually reserved for mid-level rotations within severe thunderstorms, and are warm core cyclones driven by latent heat of its associated thunderstorm activity.

Tornadoes form in the warm sector of extratropical cyclones where a strong upper level jet stream exists. Mesocyclones are believed to form when strong changes of wind speed and/or direction with height ("wind shear") sets parts of the lower part of the atmosphere spinning in invisible tube-like rolls. The convective updraft of a thunderstorm is then thought to draw up this spinning air, tilting the rolls' orientation upward (from parallel to the ground to perpendicular) and causing the entire updraft to rotate as a vertical column.

As the updraft rotates, it may form what is known as a wall cloud. The wall cloud is a spinning layer of clouds descending from the mesocyclone. The wall cloud tends to form closer to the center of the mesocyclone. It should be noted the wall clouds do not necessarily need a mesocyclone to form and do not always rotate. As the wall cloud descends, a funnel-shaped cloud may form at its center.

This is the first stage of tornado formation. The presence of a mesocyclone is believed to be a key factor in the formation of the strong tornadoes associated with severe thunderstorms.

Tornadoes

Tornadoes exist on the microscale or low end of the mesoscale gamma domain. The cycle begins when a strong thunderstorm develops a rotating mesocyclone a few miles up in the atmosphere, becoming a supercell. As rainfall in the storm increases, it drags with it an area of quickly descending air known as the rear flank downdraft (RFD). This downdraft accelerates as it approaches the ground, and drags the rotating mesocyclone towards the ground with it.

As the mesocyclone approaches the ground, a visible condensation funnel appears to descend from the base of the storm, often from a rotating wall cloud. As the funnel descends, the RFD also reaches the ground, creating a gust front that can cause damage a good distance from the tornado. Usually, the funnel cloud begins causing damage on the ground (becoming a tornado) within minutes of the RFD reaching the ground.

Waterspouts

Waterspouts exist on the microscale. While some waterspouts are strong (tornadic) like their land-based counterparts, most are much weaker and caused by different atmospheric dynamics. They normally develop in moisture-laden environments with little vertical wind shear along lines of convergence, such as land breezes, lines of frictional convergence from nearby landmasses, or surface troughs. Their parent cloud can be as innocuous as a moderate cumulus, or as significant as a thunderstorm. Waterspouts normally develop as their parent clouds are in the process of development, and it is theorized that they spin up as they move up the surface boundary from the horizontal wind shear near the surface, and then stretch upwards to the cloud once the low level shear vortex aligns with a developing cumulus or thunderstorm. Weak tornadoes, known as landspouts, across eastern Colorado have been witnessed to develop in a similar manner. An outbreak occurred in the Great Lakes in late September and early October 2003 along a lake effect band. September is the peak month of landspout and waterspout occurrence around Florida and for waterspout occurrence around the Great Lakes.

Related Terms

Cyclogenesis is the opposite of cyclolysis, which concerns the weakening of surface cyclones. The term has an anticyclonic (high pressure system) equivalent—Anticyclogenesis, which deals with the formation of surface high pressure systems.

Storm

A storm is any disturbed state of an environment or astronomical body's atmosphere especially affecting its surface, and strongly implying severe weather. It may be marked by significant disruptions to normal conditions such as strong wind, hail, thunder and lightning (a thunderstorm), heavy precipitation (snowstorm, rainstorm), heavy freezing rain (ice storm), strong winds (tropical cyclone, windstorm), or wind transporting some substance through the atmosphere as in a dust storm, blizzard, sandstorm, etc.

Storms have the potential to harm lives and property via storm surge, heavy rain or snow causing flooding or road impassibility, lightning, wildfires, and vertical wind shear, however, systems with significant rainfall and duration help alleviate drought in places they move through. Heavy snowfall can allow special recreational activities to take place which would not be possible otherwise, such as skiing and snowmobiling.

Desert storms are often accompanied by violent winds, and pass rapidly.

The English word comes from Proto-Germanic *sturmaz* meaning "noise, tumult".

Lightning storm, Port-la-Nouvelle.

Formation

Satellite image of the intense nor'easter responsible for the North American blizzard of 2006. Note the hurricane-like eye at the center.

Storms are created when a center of low pressure develops with a system of high pressure surrounding it. This combination of opposing forces can create winds and result in the formation of storm clouds, such as the cumulonimbus. Small localized areas of low pressure can form from hot air rising off hot ground, resulting in smaller disturbances such as dust devils and whirlwinds.

Types

Classic storm of summer, in Sierras de Córdoba, Argentina.

There are many varieties and names for storms:

- Ice storm — Ice storms are one of the most dangerous forms of winter storms. When surface temperatures are below freezing, but a thick layer of above-freezing air remains aloft, rain can fall into the freezing layer and freeze upon impact into a glaze of ice. In general, 8 millimetres (0.31 in) of accumulation is all that is required, especially in combination with breezy conditions, to start downing power lines as well as tree limbs. Ice storms also make unheated road surfaces too slick to drive upon. Ice storms can vary in time range from hours to days and can cripple small towns and large urban centers alike.

- Blizzard — There are varying definitions for blizzards, both over time and by location. In general, a blizzard is accompanied by gale-force winds, heavy snow (accumulating at a rate of at least 5 centimeters (2 in) per hour), and very cold conditions (below approximately -10 degrees Celsius or 14 F). Lately, the temperature criterion has fallen out of the definition across the United States

- Snowstorm — A heavy fall of snow accumulating at a rate of more than 5 centimeters (2 in) per hour that lasts several hours. Snow storms, especially ones with a high liquid equivalent and breezy conditions, can down tree limbs, cut off power, and paralyze travel over a large region.

- Ocean Storm — Storm conditions out at sea are defined as having sustained winds of 48 knots (55 mph or 90 km/h) or greater. Usually just referred to as a storm, these systems can sink vessels of all types and sizes.

- Firestorm — Firestorms are conflagrations which attain such intensity that they create and sustain their own wind systems. It is most commonly a natural phenomenon, created during some of the largest bushfires, forest fires, and wildfires. The Peshtigo Fire is one example of a firestorm. Firestorms can also be deliberate effects of targeted explosives such as occurred as a result of the aerial bombings of Dresden. Nuclear detonations generate firestorms if high winds are not present.

- Dust devil — a small, localized updraft of rising air.

- Wind storm— A storm marked by high wind with little or no precipitation. Windstorm damage often opens the door for massive amounts of water and debris to cause further damage to a structure. European windstorms and derechos are two type of windstorms. High Wind is also the cause of Sandstorms in dry climates.

- Squall — sudden onset of wind increase of at least 16 knots (30 km/h) or greater sustained for at least one minute.

- Gale — An extratropical storm with sustained winds between 34-48 knots (39-55 mph or 63–90 km/h).

- Thunderstorm — A thunderstorm is a type of storm that generates lightning and the attendant thunder. It is normally accompanied by heavy precipitation. Thunderstorms occur throughout the world, with the highest frequency in tropical rainforest regions where there are conditions of high humidity and temperature along with atmospheric instability. These storms occur when high levels of condensation form in a volume of unstable air that generates deep, rapid, upward motion in the atmosphere. The heat energy creates powerful rising air currents that swirl upwards to the tropopause. Cool descending air currents produce strong downdraughts below the storm. After the storm has spent its energy, the rising currents die away and downdraughts break up the cloud. Individual storm clouds can measure 2–10 km across.

- Tropical cyclone — A tropical cyclone is a storm system with a closed circulation around a centre of low pressure, fueled by the heat released when moist air rises and condenses. The name underscores its origin in the tropics and their cyclonic nature. Tropical cyclones are distinguished from other cyclonic storms such as nor'easters and polar lows by the heat mechanism that fuels them, which makes them "warm core" storm systems.

  Tropical cyclones form in the oceans if the conditions in the area are favorable, and depending on their strength and location, there are various terms by which they are called, such as *tropical depression*, *tropical storm*, *hurricane* and *typhoon*.

- Hailstorm — a type of storm that precipitates round chunks of ice. Hailstorms usually occur during regular thunder storms. While most of the hail that precipitates from the clouds is fairly small and virtually harmless, there are occasional occurrences of hail greater than 2 inches (5 cm) in diameter that can cause much damage and injuries.

- Tornado — A tornado is a violent, destructive wind storm occurring on land. Usually its appearance is that of a dark, funnel-shaped cloud. Often tornadoes are preceded by thunderstorms and a wall cloud. They are often called the most destructive of storms, and while they form all over the world, the interior of the United States is the most prone area, especially throughout Tornado Alley.

A tornado in Binger, Oklahoma during the 1981 outbreak.

Classification

A strict meteorological definition of a terrestrial storm is a wind measuring 10 or higher on the Beaufort scale, meaning a wind speed of 24.5 m/s (89 km/h, 55 mph) or more; however, popular usage is not so restrictive. Storms can last anywhere from 12 to 200 hours, depending on season and geography. In North America, the east and northeast storms are noted for the most frequent repeatability and duration, especially during the cold period. Big terrestrial storms alter the ocean-ographic conditions that in turn may affect food abundance and distribution: strong currents, strong tides, increased siltation, change in water temperatures, overturn in the water column, etc.

Extraterrestrial Storms

The Great Red Spot on Jupiter

Storms do not only occur on Earth; other planetary bodies with a sufficient atmosphere (gas giants in particular) also undergo stormy weather. The Great Red Spot on Jupiter provides a well-known example. Though technically an anticyclone with greater than hurricane wind speeds, it is larger than the Earth and has persisted for at least 340 years, having first been observed by astronomer Galileo Galilei. Neptune also had its own lesser-known Great Dark Spot.

In September 1994 the Hubble telescope – using Wide Field Planetary Camera 2 – imaged storms on Saturn generated by upwelling of warmer air, similar to a terrestrial thunderhead. The east-west extent of the same-year storm equalled the diameter of Earth. The storm was observed earlier in September 1990 and acquired the name Dragon Storm.

The dust storms of Mars vary in size, but can often cover the entire planet. They tend to occur when Mars comes closest to the Sun, and have been shown to increase the global temperature.

One particularly large Martian storm was exhaustively studied up close due to coincidental timing. When the first spacecraft to successfully orbit another planet, Mariner 9, arrived and successfully orbited Mars on 14 November 1971, planetary scientists were surprised to find the atmosphere was thick with a planet-wide robe of dust, the largest storm ever observed on Mars. The surface of the planet was totally obscured. Mariner 9's computer was reprogrammed from Earth to delay imaging of the surface for a couple of months until the dust settled. However, the surface-obscured images contributed much to the collection of Mars atmospheric and planetary surface science.

Two extrasolar planets are known to have storms: HD 209458 b and HD 80606 b. The former's storm was discovered on June 23, 2010 and measured at 6,200 km/h, while the latter produces winds of 17,700 kilometers (11,000 mi) per hour across the surface. The spin of the planet then creates giant swirling shock-wave storms that carry the heat aloft.

Effects on Human Society

A snow blockade in southern Minnesota, US in 1881

Shipwrecks are common with the passage of strong tropical cyclones. Such shipwrecks can change the course of history, as well as influence art and literature. A hurricane led to a victory of the Spanish over the French for control of Fort Caroline, and ultimately the Atlantic coast of North America, in 1565.

Strong winds from any storm type can damage or destroy vehicles, buildings, bridges, and other outside objects, turning loose debris into deadly flying projectiles. In the United States, major hurricanes comprise just 21% of all landfalling tropical cyclones, but account for 83% of all damage. Tropical cyclones often knock out power to tens or hundreds of thousands of people, preventing vital communication and hampering rescue efforts. Tropical cyclones often destroy key bridges, overpasses, and roads, complicating efforts to transport food, clean water, and medicine to the areas that need it. Furthermore, the damage caused by tropical cyclones to buildings and dwellings can result in economic damage to a region, and to a diaspora of the population of the region.

A return stroke, cloud-to-ground lightning strike during a thunderstorm.

The storm surge, or the increase in sea level due to the cyclone, is typically the worst effect from landfalling tropical cyclones, historically resulting in 90% of tropical cyclone deaths. The relatively quick surge in sea level can move miles/kilometers inland, flooding homes and cutting off escape routes. The storm surges and winds of hurricanes may be destructive to human-made structures, but they also stir up the waters of coastal estuaries, which are typically important fish breeding locales.

Cloud-to-ground lightning frequently occurs within the phenomena of thunderstorms and have numerous hazards towards landscapes and populations. One of the more significant hazards lightning can pose is the wildfires they are capable of igniting. Under a regime of low precipitation (LP) thunderstorms, where little precipitation is present, rainfall cannot prevent fires from starting when vegetation is dry as lightning produces a concentrated amount of extreme heat. Wildfires can devastate vegetation and the biodiversity of an ecosystem. Wildfires that occur close to urban environments can inflict damages upon infrastructures, buildings, crops, and provide risks to explosions, should the flames be exposed to gas pipes. Direct damage caused by lightning strikes occurs on occasion. In areas with a high frequency for cloud-to-ground lightning, like Florida, lightning causes several fatalities per year, most commonly to people working outside.

Precipitation with low potential of hydrogen levels (pH), otherwise known as acid rain, is also a frequent risk produced by lightning. Distilled water, which contains no carbon dioxide, has a neutral pH of 7. Liquids with a pH less than 7 are acidic, and those with a pH greater than 7 are bases. "Clean" or unpolluted rain has a slightly acidic pH of about 5.2, because carbon dioxide and water in the air react together to form carbonic acid, a weak acid (pH 5.6 in distilled water), but unpolluted rain also contains other chemicals. Nitric oxide present during thunderstorm phenomena, caused by the splitting of nitrogen molecules, can result in the production of acid rain, if nitric oxide forms compounds with the water molecules in precipitation, thus creating acid rain. Acid rain can damage infrastructures containing calcite or other solid chemical compounds containing carbon. In ecosystems, acid rain can dissolve plant tissues of vegetations and increase acidification process in bodies of water and in soil, resulting in deaths of marine and terrestrial organisms.

Hail damage to roofs often goes unnoticed until further structural damage is seen, such as leaks or cracks. It is hardest to recognize hail damage on shingled roofs and flat roofs, but all roofs have their own hail damage detection problems. Metal roofs are fairly resistant to hail damage, but may accumulate cosmetic damage in the form of dents and damaged coatings. Hail is also a common nuisance to drivers of automobiles, severely denting the vehicle and cracking or even shattering windshields and windows. Rarely, massive hailstones have been known to cause concussions or fatal head trauma. Hailstorms have been the cause of costly and deadly events throughout history. One of the earliest recorded incidents occurred around the 9th century in Roopkund, Uttarakhand, India. The largest hailstone in terms of diameter and weight ever recorded in the United States fell on July 23, 2010 in Vivian, South Dakota in the United States; it measured 8 inches (20 cm) in diameter and 18.62 inches (47.3 cm) in circumference, weighing in at 1.93 pounds (0.88 kg). This broke the previous record for diameter set by a hailstone 7 inches diameter and 18.75 inches circumference which fell in Aurora, Nebraska in the United States on June 22, 2003, as well as the record for weight, set by a hailstone of 1.67 pounds (0.76 kg) that fell in Coffeyville, Kansas in 1970.

Various hazards, ranging from hail to lightning can affect outside technology facilities, such as antennas, satellite dishes, and towers. As a result, companies with outside facilities have begun installing such facilities underground, in order to reduce the risk of damage from storms.

Substantial snowfall can disrupt public infrastructure and services, slowing human activity even in regions that are accustomed to such weather. Air and ground transport may be greatly inhibited or shut down entirely. Populations living in snow-prone areas have developed various ways to travel across the snow, such as skis, snowshoes, and sleds pulled by horses, dogs, or other animals and later, snowmobiles. Basic utilities such as electricity, telephone lines, and gas supply can also fail. In addition, snow can make roads much harder to travel and vehicles attempting to use them can easily become stuck.

The combined effects can lead to a "snow day" on which gatherings such as school, work, or church are officially canceled. In areas that normally have very little or no snow, a snow day may occur when there is only light accumulation or even the threat of snowfall, since those areas are unprepared to handle any amount of snow. In some areas, such as some states in the United States, schools are given a yearly quota of snow days (or "calamity days"). Once the quota is exceeded, the snow days must be made up. In other states, all snow days must be made up. For example, schools may extend the remaining school days later into the afternoon, shorten spring break, or delay the start of summer vacation.

Accumulated snow is removed to make travel easier and safer, and to decrease the long-term effect of a heavy snowfall. This process utilizes shovels and snowplows, and is often assisted by sprinkling salt or other chloride-based chemicals, which reduce the melting temperature of snow. In some areas with abundant snowfall, such as Yamagata Prefecture, Japan, people harvest snow and store it surrounded by insulation in ice houses. This allows the snow to be used through the summer for refrigeration and air conditioning, which requires far less electricity than traditional cooling methods.

Agriculture

Hail can cause serious damage, notably to automobiles, aircraft, skylights, glass-roofed structures, livestock, and most commonly, farmers' crops. Wheat, corn, soybeans, and tobacco are the most sensitive crops to hail damage. Hail is one of Canada's most expensive hazards. Snowfall can be beneficial to agriculture by serving as a thermal insulator, conserving the heat of the Earth and protecting crops from subfreezing weather. Some agricultural areas depend on an accumulation of snow during winter that will melt gradually in spring, providing water for crop growth. If it melts into water and refreezes upon sensitive crops, such as oranges, the resulting ice will protect the fruit from exposure to lower temperatures. Although tropical cyclones take an enormous toll in lives and personal property, they may be important factors in the precipitation regimes of places they affect and bring much-needed precipitation to otherwise dry regions. Hurricanes in the eastern north Pacific often supply moisture to the Southwestern United States and parts of Mexico. Japan receives over half of its rainfall from typhoons. Hurricane Camille averted drought conditions and ended water deficits along much of its path, though it also killed 259 people and caused $9.14 billion (2005 USD) in damage.

Aviation

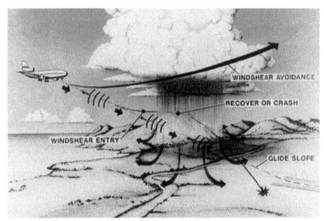

Effect of wind shear on aircraft trajectory. Merely correcting for the initial gust front can have dire consequences.

Hail is one of the most significant thunderstorm hazards to aircraft. When hail stones exceed 0.5 inches (13 mm) in diameter, planes can be seriously damaged within seconds. The hailstones accumulating on the ground can also be hazardous to landing aircraft. Strong wind outflow from thunderstorms causes rapid changes in the three-dimensional wind velocity just above ground level. Initially, this outflow causes a headwind that increases airspeed, which normally causes a pilot to reduce engine power if they are unaware of the wind shear. As the aircraft passes into the region of the downdraft, the localized headwind diminishes, reducing the aircraft's airspeed and increasing its sink rate. Then, when the aircraft passes through the other side of the downdraft, the headwind becomes a tailwind, reducing lift generated by the wings, and leaving the aircraft in a low-power, low-speed descent. This can lead to an accident if the aircraft is too low to effect a recovery before ground contact. As the result of the accidents in the 1970s and 1980s, in 1988 the U.S. Federal Aviation Administration mandated that all commercial aircraft have on-board wind shear detection systems by 1993. Between 1964 and 1985, wind shear directly caused or contributed to 26 major civil transport aircraft accidents in the U.S. that led to 620 deaths and 200 injuries. Since 1995, the number of major civil aircraft accidents caused by wind shear has dropped to approximately one every ten years, due to the mandated on-board detection as well as the addition of Doppler weather radar units on the ground. (NEXRAD)

Recreation

Many winter sports, such as skiing, snowboarding, snowmobiling, and snowshoeing depend upon snow. Where snow is scarce but the temperature is low enough, snow cannons may be used to produce an adequate amount for such sports. Children and adults can play on a sled or ride in a sleigh. Although a person's footsteps remain a visible lifeline within a snow-covered landscape, snow cover is considered a general danger to hiking since the snow obscures landmarks and makes the landscape itself appear uniform.

Notable Storms in Art and Culture

The Great Wave off Kanagawa, a Ukiyo-e print by Hokusai.

In Mythology and Literature

According to the Bible, a giant storm sent by God flooded the Earth. Noah and his family and the animals entered the Ark, and "the same day were all the fountains of the great deep broken up, and the windows of heaven were opened, and the rain was upon the earth forty days and forty nights." The flood covered even the highest mountains to a depth of more than twenty feet, and all creatures died; only Noah and those with him on the Ark were left alive. In the New Testament, Jesus Christ is recorded to have calmed a storm on the Sea of Galilee.

The Gilgamesh flood myth is a deluge story in the *Epic of Gilgamesh*.

In Greek mythology there were several gods of storms: Briareos, the god of sea storms; Aigaios, a god of the violent sea storms; and Aiolos, keeper of storm-winds, squalls and tempests.

The *Sea Venture* was wrecked near Bermuda in 1609, which led to the colonization of Bermuda and provided the inspiration for Shakespeare's play *The Tempest*(1611). Specifically, Sir Thomas Gates, future governor of Virginia, was on his way to England from Jamestown, Virginia. On Saint James Day, while he was between Cuba and the Bahamas, a hurricane raged for nearly two days. Though one of the small vessels in the fleet sank to the bottom of the Florida Straits, seven of the remaining vessels reached Virginia within several days after the storm. The flagship of the fleet, known as *Sea Adventure*, disappeared and was presumed lost. A small bit of fortune befell the ship and her crew when they made landfall on Bermuda. The vessel was damaged on a surrounding coral reef, but all aboard survived for nearly a year on the island. The British colonists claimed the island and quickly settled Bermuda. In May 1610, they set forth for Jamestown, this time arriving at their destination.

The children's novel *The Wonderful Wizard of Oz*, written by L. Frank Baum and illustrated by W. W. Denslow, chronicles the adventures of a young girl named Dorothy Gale in the Land of Oz, after being swept away from her Kansas farm home by a tornado. The story was originally published by the George M. Hill Company in Chicago on May 17, 1900 and has since been reprinted numerous times, most often under the name *The Wizard of Oz*, and adapted for use in other media. Thanks in part to the 1939 MGM movie, it is one of the best-known stories in American popular culture and has been widely translated. Its initial success, and the success of the popular 1902 Broadway musical which Baum adapted from his original story, led to Baum's writing thirteen more Oz books.

Hollywood director King Vidor (February 8, 1894 – November 1, 1982) survived the Galveston Hurricane of 1900 as a boy. Based on that experience, he published a fictionalized account of that cyclone, titled "Southern Storm", for the May 1935 issue of *Esquire* magazine. Erik Larson excerpts a passage from that article in his 2005 book, *Isaac's Storm*:

> I remember now that it seemed as if we were in a bowl looking up toward the level of the sea. As we stood there in the sandy street, my mother and I, I wanted to take my mother's hand and hurry her away. I felt as if the sea was going to break over the edge of the bowl and come puring down upon us.

Numerous other accounts of the Galveston Hurricane of 1900 have been made in print and in film. Larson cites many of them in *Isaac's Storm*, which centrally features that storm, as well as chronicles the creation of the Weather Bureau (which came to known as the National Weather Service) and that agency's fateful rivalry with the weather service in Cuba, and a number of other major storms, such as those which ravaged Indianola, Texas in 1875 and 1886.

The Great Storm of 1987 is key in an important scene near the end of *Possession: A Romance*, the bestselling and Man Booker Prize-winning novel by A. S. Byatt. The Great Storm of 1987 occurred on the night of October 15–16, 1987, when an unusually strong weather system caused winds to hit much of southern England and northern France. It was the worst storm to hit England since the Great Storm of 1703 (284 years earlier) and was responsible for the deaths of at least 22 people in England and France combined (18 in England, at least four in France).

Hurricane Katrina (2005) has been featured in a number of works of fiction.

In Fine Art

The Romantic seascape painters J. M. W. Turner and Ivan Aivazovsky created some of the most lasting impressions of the sublime and stormy seas that are firmly imprinted on the popular mind. Turner's representations of powerful natural forces reinvented the traditional seascape during the first half of the nineteenth century.

Upon his travels to Holland, he took note of the familiar large rolling waves of the English seashore transforming into the sharper, choppy waves of a Dutch storm. A characteristic example of

Turner's dramatic seascape is *The Slave Ship* of 1840. Aivazovsky left several thousand turbulent canvases in which he increasingly eliminated human figures and historical background to focus on such essential elements as light, sea, and sky. His grandiose *Ninth Wave* (1850) is an ode to human daring in the face of the elements.

Rembrandt's *The Storm on the Sea of Galilee.*

In Motion Pictures

The 1926 silent film *The Johnstown Flood* features the Great Flood of 1889 in Johnstown, Pennsylvania. The flood, caused by the catastrophic failure of the South Fork Dam after days of extremely heavy rainfall, prompted the first major disaster relief effort by the American Red Cross, directed by Clara Barton. The Johnstown Flood was depicted in numerous other media (both fictional and in non-fiction), as well.

Warner Bros.' 2000 dramatic disaster film *The Perfect Storm*, directed by Wolfgang Petersen, is an adaptation of Sebastian Junger's 1997 non-fiction book of the same title. The book and film feature the crew of the *Andrea Gail*, which got caught in the Perfect Storm of 1991. The 1991 Perfect Storm, also known as the Halloween Nor'easter of 1991, was a nor'easter that absorbed Hurricane Grace and ultimately evolved into a small hurricane late in its life cycle.

In Music

Storms were also portrayed in several works of music. Examples are Beethoven's *Pastoral Symphony* (the fourth movement), Presto of the violin concerto RV 315 (*Summer*) from the *Four Seasons* by Vivaldi, and a scene in Act II of Rossini's opera *The Barber of Seville.*

Gallery

Lightning within the cloud causes the entire blanket to illuminate.

High Desert storm approaches at sunset.

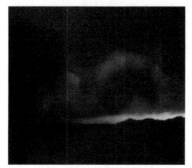

Heavy storm brought by Severe Tropical Storm Sanvu in Hong Kong.

References

- World Meteorological Organization, ed. (1975). International Cloud Atlas, preface to the 1939 edition. (PDF). I. pp. IX–XIII. ISBN 92-63-10407-7. Retrieved 6 December 2014.

- World Meteorological Organization, ed. (1975). Nacreous, International Cloud Atlas (PDF). I. p. 65. ISBN 92-63-10407-7. Retrieved 26 August

- World Meteorological Organization, ed. (1975). Cirrostratus, International Cloud Atlas (PDF). I. pp. 29–31. ISBN 92-63-10407-7. Retrieved 26 August 2014.

- Miyazaki, R.; Yoshida, S.; Dobashi, Y.; Nishita, T. (2001). "A method for modeling clouds based on atmospheric fluid dynamics". Proceedings Ninth Pacific Conference on Computer Graphics and Applications. Pacific Graphics 2001. p. 363. doi:10.1109/PCCGA.2001.962893. ISBN 0-7695-1227-5.

- World Meteorological Organization, ed. (1975). Altostratus, International Cloud Atlas (PDF). I. pp. 35–37. ISBN 92-63-10407-7. Retrieved 26 August 2014.

- Lee M. Grenci; Jon M. Nese (2001). A World of Weather: Fundamentals of Meteorology: A Text / Laboratory Manual (3 ed.). Kendall/Hunt Publishing Company. pp. 207–212. ISBN 978-0-7872-7716-1. OCLC 51160155.

- Pretor-Pinney, Gavin (2007). The Cloudspotter's Guide: The Science, History, and Culture of Clouds. Penguin Group. p. 20. ISBN 978-1-101-20331-6.

- Kondratev, Kirill Iakovlevich (2006). Atmospheric aerosol properties: formation, processes and impacts. Springer. p. 403. ISBN 978-3-540-26263-3.

- Bougher, Stephen Wesley; Phillips, Roger (1997). Venus II: Geology, Geophysics, Atmosphere, and Solar Wind Environment. University of Arizona Press. pp. 127–129. ISBN 978-0-8165-1830-2.

- Dougherty, Michele; Esposito, Larry (November 2009). Saturn from Cassini-Huygens (1 ed.). Springer. p. 118. ISBN 978-1-4020-9216-9. OCLC 527635272

- Ludlum, David McWilliams (2000). National Audubon Society Field Guide to Weather. Alfred A. Knopf. p. 473. ISBN 0-679-40851-7. OCLC 56559729.

- Klein, Alice; Keenan, Greta (8 July 2016). "Perfect storm hits Taiwan as China sees worst floods in 20 years". New Scientist. Retrieved 10 July 2016.

- Colorado State University Dept. of Atmospheric Science, ed. (2015). "Cloud type identification by satellites" (PDF). Colorado State University. Retrieved 30 December 2015.

- Vincent J. Schaefer (October 1952). "Cloud Forms of the Jet Stream". Tellus. General Electric Research Laboratory. 5: 27–31. doi:10.1111/j.2153-3490.1953.tb01032.x. Retrieved 27 November 2014.

- Task Team On Revision of the International Cloud Atlas (2013). "Final Report, pp 15–18" (PDF). World Meteorological Organization. Retrieved 6 October 2014.

- Long, Michael J.; Hanks, Howard H.; Beebe, Robert G. (June 1965). "TROPOPAUSE PENETRATIONS BY CUMULONIMBUS CLOUDS". Retrieved 9 November 2014.

An Integrated Study of Weather Forecasting

This chapter focuses on the science of weather forecasting which helps predict the state of the atmosphere for a given region. It explains how weather forecasts are made and the tools, techniques and technology used for it. Listed here are topics like weather forecasting and weather map. There is a section devoted to surface weather analysis as well.

Weather Forecasting

Weather forecasting is the application of science and technology to predict the state of the atmosphere for a given location. Human beings have attempted to predict the weather informally for millennia, and formally since the nineteenth century. Weather forecasts are made by collecting quantitative data about the current state of the atmosphere at a given place and using scientific understanding of atmospheric processes to project how the atmosphere will change.

Once an all-human endeavor based mainly upon changes in barometric pressure, current weather conditions, and sky condition, weather forecasting now relies on computer-based models that take many atmospheric factors into account. Human input is still required to pick the best possible forecast model to base the forecast upon, which involves pattern recognition skills, teleconnections, knowledge of model performance, and knowledge of model biases. The inaccuracy of forecasting is due to the chaotic nature of the atmosphere, the massive computational power required to solve the equations that describe the atmosphere, the error involved in measuring the initial conditions, and an incomplete understanding of atmospheric processes. Hence, forecasts become less accurate as the difference between current time and the time for which the forecast is being made (the *range* of the forecast) increases. The use of ensembles and model consensus help narrow the error and pick the most likely outcome.

There are a variety of end uses to weather forecasts. Weather warnings are important forecasts because they are used to protect life and property. Forecasts based on temperature and precipitation are important to agriculture, and therefore to traders within commodity markets. Temperature forecasts are used by utility companies to estimate demand over coming days. On an everyday basis, people use weather forecasts to determine what to wear on a given day. Since outdoor activities are severely curtailed by heavy rain, snow and wind chill, forecasts can be used to plan activities around these events, and to plan ahead and survive them.

History

Ancient Forecasting

For millennia people have tried to forecast the weather. In 650 BC, the Babylonians predicted the weather from cloud patterns as well as astrology. In about 340 BC, Aristotle described weather patterns in *Meteorologica*. Later, Theophrastus compiled a book on weather forecasting, called the *Book of Signs*. Chinese weather prediction lore extends at least as far back as 300 BC, which was also around the same time ancient Indian astronomers developed weather-prediction methods. In 904 AD, Ibn Wahshiyya's *Nabatean Agriculture* discussed the weather forecasting of atmospheric changes and signs from the planetary astral alterations; signs of rain based on observation of the lunar phases; and weather forecasts based on the movement of winds.

Ancient weather forecasting methods usually relied on observed patterns of events, also termed pattern recognition. For example, it might be observed that if the sunset was particularly red, the following day often brought fair weather. This experience accumulated over the generations to produce weather lore. However, not all of these predictions prove reliable, and many of them have since been found not to stand up to rigorous statistical testing.

Modern Methods

The *Royal Charter* sank in an 1859 storm, stimulating the establishment of modern weather forecasting.

It was not until the invention of the electric telegraph in 1835 that the modern age of weather forecasting began. Before that, the fastest that distant weather reports could travel was around 100 miles per day (160 km/d), but was more typically 40–75 miles per day (60–120 km/day) (whether by land or by sea). By the late 1840s, the telegraph allowed reports of weather conditions from a wide area to be received almost instantaneously, allowing forecasts to be made from knowledge of weather conditions further upwind.

The two men credited with the birth of forecasting as a science were officer of the Royal Navy Francis Beaufort and his protégé Robert FitzRoy. Both were influential men in British naval and governmental circles, and though ridiculed in the press at the time, their work gained scientific credence, was accepted by the Royal Navy, and formed the basis for all of today's weather forecasting knowledge.

Beaufort developed the Wind Force Scale and Weather Notation coding, which he was to use in his journals for the remainder of his life. He also promoted the development of reliable tide tables around British shores, and with his friend William Whewell, expanded weather record-keeping at 200 British Coast guard stations.

Robert FitzRoy was appointed in 1854 as chief of a new department within the Board of Trade to deal with the collection of weather data at sea as a service to mariners. This was the forerunner of the modern Meteorological Office. All ship captains were tasked with collating data on the weather and computing it, with the use of tested instruments that were loaned for this purpose.

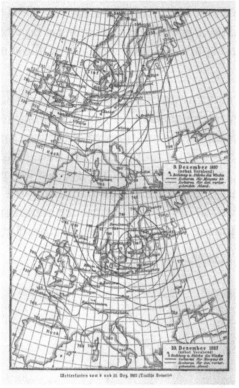

Weather map of Europe, 10 December 1887.

A storm in 1859 that caused the loss of the *Royal Charter* inspired FitzRoy to develop charts to allow predictions to be made, which he called *"forecasting the weather"*, thus coining the term "weather forecast". Fifteen land stations were established to use the new telegraph to transmit to him daily reports of weather at set times leading to the first gale warning service. His warning service for shipping was initiated in February 1861, with the use of telegraph communications. The first daily weather forecasts were published in *The Times* in 1861. In the following year a system was introduced of hoisting storm warning cones at the principal ports when a gale was expected. The *"Weather Book"* which FitzRoy published in 1863 was far in advance of the scientific opinion of the time.

As the electric telegraph network expanded, allowing for the more rapid dissemination of warnings, a national observational network was developed which could then be used to provide synoptic analyses. Instruments to continuously record variations in meteorological parameters using photography were supplied to the observing stations from Kew Observatory – these cameras had been invented by Francis Ronalds in 1845 and his barograph had earlier been used by FitzRoy.

To convey accurate information, it soon became necessary to have a standard vocabulary describing clouds; this was achieved by means of a series of classifications first achieved by Luke Howard in 1802, and standardized in the *International Cloud Atlas* of 1896.

Numerical Prediction

It was not until the 20th century that advances in the understanding of atmospheric physics led to the foundation of modern numerical weather prediction. In 1922, English scientist Lewis Fry Richardson published "Weather Prediction By Numerical Process", after finding notes and derivations he worked on as an ambulance driver in World War I. He described therein how small terms in the prognostic fluid dynamics equations governing atmospheric flow could be neglected, and a finite differencing scheme in time and space could be devised, to allow numerical prediction solutions to be found.

Richardson envisioned a large auditorium of thousands of people performing the calculations and passing them to others. However, the sheer number of calculations required was too large to be completed without the use of computers, and the size of the grid and time steps led to unrealistic results in deepening systems. It was later found, through numerical analysis, that this was due to numerical instability. The first computerised weather forecast was performed by a team led by the mathematician John von Neumann; von Neumann publishing the paper *Numerical Integration of the Barotropic Vorticity Equation* in 1950. Practical use of numerical weather prediction began in 1955, spurred by the development of programmable electronic computers.

Broadcasts

George Cowling (above) presented the first in-vision forecast on 11 January 1954 for the BBC.

The first ever daily weather forecasts were published in *The Times* on 1 August 1861, and the first weather maps were produced later in the same year. In 1911, the Met Office began issuing the first marine weather forecasts via radio transmission. These included gale and storm warnings for areas around Great Britain. In the United States, the first public radio forecasts were made in 1925 by Edward B. "E.B." Rideout, on WEEI, the Edison Electric Illuminating station in Boston. Rideout came from the U.S. Weather Bureau, as did WBZ weather forecaster G. Harold Noyes in 1931.

The world's first televised weather forecasts, including the use of weather maps, were experimentally broadcast by the BBC in 1936. This was brought into practice in 1949 after World War II. George Cowling gave the first weather forecast while being televised in front of the map in 1954. In America, experimental television forecasts were made by James C Fidler in Cincinnati in either 1940 or 1947 on the DuMont Television Network. In the late 1970s and early 80s, John Coleman, the first weatherman on ABC-TV's Good Morning America, pioneered the use of on-screen weather satellite information and computer graphics for television forecasts. Coleman was a co-founder of The Weather Channel (TWC) in 1982. TWC is now a 24-hour cable network.

How Models Create Forecasts

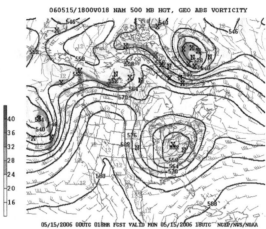

An example of 500 mbar geopotential height and absolute vorticity prediction
from a numerical weather prediction model

The basic idea of numerical weather prediction is to sample the state of the fluid at a given time and use the equations of fluid dynamics and thermodynamics to estimate the state of the fluid at some time in the future. The main inputs from country-based weather services are surface observations from automated weather stations at ground level over land and from weather buoys at sea. The World Meteorological Organization acts to standardize the instrumentation, observing practices and timing of these observations worldwide. Stations either report hourly in METAR reports, or every six hours in SYNOP reports. Sites launch radiosondes, which rise through the depth of the troposphere and well into the stratosphere. Data from weather satellites are used in areas where traditional data sources are not available. Compared with similar data from radiosondes, the satellite data has the advantage of global coverage, however at a lower accuracy and resolution. Meteorological radar provide information on precipitation location and intensity, which can be used to estimate precipitation accumulations over time. Additionally, if a pulse Doppler weather radar is used then wind speed and direction can be determined.

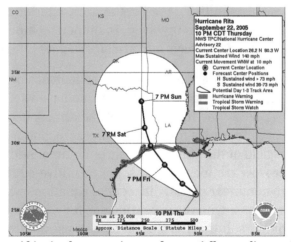

Modern weather predictions aid in timely evacuations and potentially save lives and prevent property damage

Commerce provides pilot reports along aircraft routes, and ship reports along shipping routes. Research flights using reconnaissance aircraft fly in and around weather systems of interest such as tropical cyclones. Reconnaissance aircraft are also flown over the open oceans during the cold season into systems which cause significant uncertainty in forecast guidance, or are expected to be of high impact 3–7 days into the future over the downstream continent.

Models are *initialized* using this observed data. The irregularly spaced observations are processed by data assimilation and objective analysis methods, which perform quality control and obtain values at locations usable by the model's mathematical algorithms (usually an evenly spaced grid). The data are then used in the model as the starting point for a forecast. Commonly, the set of equations used to predict the known as the physics and dynamics of the atmosphere are called primitive equations. These equations are initialized from the analysis data and rates of change are determined. The rates of change predict the state of the atmosphere a short time into the future. The equations are then applied to this new atmospheric state to find new rates of change, and these new rates of change predict the atmosphere at a yet further time into the future. This *time stepping* procedure is continually repeated until the solution reaches the desired forecast time. The length of the time step is related to the distance between the points on the computational grid.

The length of the time step chosen within the model is related to the distance between the points on the computational grid, and is chosen to maintain numerical stability. Time steps for global models are on the order of tens of minutes, while time steps for regional models are between one and four minutes. The global models are run at varying times into the future. The UKMET Unified Model is run six days into the future, the European Centre for Medium-Range Weather Forecasts model is run out to 10 days into the future, while the Global Forecast System model run by the Environmental Modeling Center is run 16 days into the future. The visual output produced by a model solution is known as a prognostic chart, or *prog*. The raw output is often modified before being presented as the forecast. This can be in the form of statistical techniques to remove known biases in the model, or of adjustment to take into account consensus among other numerical weather forecasts. MOS or model output statistics is a technique used to interpret numerical model output and produce site-specific guidance. This guidance is presented in coded numerical form, and can be obtained for nearly all National Weather Service reporting stations in the United States. As proposed by Edward Lorenz in 1963, long range forecasts, those made at a range of two weeks or more, are impossible to definitively predict the state of the atmosphere, owing to the chaotic nature of the fluid dynamics equations involved. In numerical models, extremely small errors in initial values double roughly every five days for variables such as temperature and wind velocity.

Essentially, a model is a computer program that produces meteorological information for future times at given locations and altitudes. Within any modern model is a set of equations, known as the primitive equations, used to predict the future state of the atmosphere. These equations—along with the ideal gas law—are used to evolve the density, pressure, and potential temperature scalar fields and the velocity vector field of the atmosphere through time. Additional transport equations for pollutants and other aerosols are included in some primitive-equation mesoscale models as well. The equations used are nonlinear partial differential equations which are impossible to solve exactly through analytical methods, with the exception of a few idealized cases. Therefore, numerical methods obtain approximate solutions. Different models use different solution methods: some global models use spectral methods for the horizontal dimensions and finite difference methods for the vertical dimension, while regional models and other global models usually use finite-difference methods in all three dimensions.

Techniques

Persistence

The simplest method of forecasting the weather, persistence, relies upon today's conditions to forecast the conditions tomorrow. This can be a valid way of forecasting the weather when it is in a steady state, such as during the summer season in the tropics. This method of forecasting strongly depends upon the presence of a stagnant weather pattern. It can be useful in both short range forecasts and long range forecasts.

Use of a Barometer

Measurements of barometric pressure and the pressure tendency (the change of pressure over time) have been used in forecasting since the late 19th century. The larger the change in pressure, especially if more than 3.5 hPa (2.6 mmHg), the larger the change in weather can be expected. If the pressure drop is rapid, a low pressure system is approaching, and there is a greater chance of rain. Rapid pressure rises are associated with improving weather conditions, such as clearing skies.

Looking at The Sky

Marestail shows moisture at high altitude, signalling the later arrival of wet weather.

Along with pressure tendency, the condition of the sky is one of the more important parameters used to forecast weather in mountainous areas. Thickening of cloud cover or the invasion of a higher cloud deck is indicative of rain in the near future. At night, high thin cirrostratus clouds can lead to halos around the moon, which indicates an approach of a warm front and its associated rain. Morning fog portends fair conditions, as rainy conditions are preceded by wind or clouds which prevent fog formation. The approach of a line of thunderstorms could indicate the approach of a cold front. Cloud-free skies are indicative of fair weather for the near future. A bar can indicate a coming tropical cyclone. The use of sky cover in weather prediction has led to various weather lore over the centuries.

Nowcasting

The forecasting of the weather within the next six hours is often referred to as nowcasting. In this time range it is possible to forecast smaller features such as individual showers and thunderstorms with reasonable accuracy, as well as other features too small to be resolved by a computer model. A human given the latest radar, satellite and observational data will be able to make a better analysis of the small scale features present and so will be able to make a more accurate forecast for the following few hours. However, there is now expert systems using those data and mesoscale numerical model to make better extrapolation, including evolution of those features in time.

Use of Forecast Models

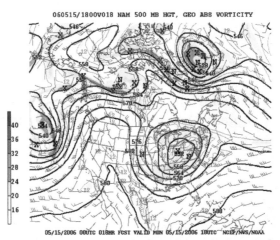

An example of 500 mbar geopotential height prediction from a numerical weather prediction model

In the past, the human forecaster was responsible for generating the entire weather forecast based upon available observations. Today, human input is generally confined to choosing a model based on various parameters, such as model biases and performance. Using a consensus of forecast models, as well as ensemble members of the various models, can help reduce forecast error. However, regardless how small the average error becomes with any individual system, large errors within any particular piece of guidance are still possible on any given model run. Humans are required to interpret the model data into weather forecasts that are understandable to the end user. Humans can use knowledge of local effects which may be too small in size to be resolved by the model to add information to the forecast. While increasing accuracy of forecast models implies that humans may no longer be needed in the forecast process at some point in the future, there is currently still a need for human intervention.

Analog Technique

The analog technique is a complex way of making a forecast, requiring the forecaster to remember a previous weather event which is expected to be mimicked by an upcoming event. What makes it a difficult technique to use is that there is rarely a perfect analog for an event in the future. Some call this type of forecasting pattern recognition. It remains a useful method of observing rainfall over data voids such as oceans, as well as the forecasting of precipitation amounts and distribution in the future. A similar technique is used in medium range forecasting, which is known as teleconnections, when systems in other locations are used to help pin down the location of another system within the surrounding regime. An example of teleconnections are by using El Niño-Southern Oscillation (ENSO) related phenomena.

Communicating Forecasts to The Public

An example of a two-day weather forecast in the visual style that an American newspaper might use. Temperatures are given in Fahrenheit.

Most end users of forecasts are members of the general public. Thunderstorms can create strong winds and dangerous lightning strikes that can lead to deaths, power outages, and widespread hail damage. Heavy snow or rain can bring transportation and commerce to a stand-still, as well as cause flooding in low-lying areas. Excessive heat or cold waves can sicken or kill those with inadequate utilities, and droughts can impact water usage and destroy vegetation.

Several countries employ government agencies to provide forecasts and watches/warnings/advisories to the public in order to protect life and property and maintain commercial interests. Knowledge of what the end user needs from a weather forecast must be taken into account to present the information in a useful and understandable way. Examples include the National Oceanic and Atmospheric Administration's National Weather Service (NWS) and Environment Canada's Meteorological Service (MSC). Traditionally, newspaper, television, and radio have been the primary outlets for presenting weather forecast information to the public. Increasingly, the internet is being used due to the vast amount of specific information that can be found. In all cases, these outlets update their forecasts on a regular basis.

Severe Weather Alerts and Advisories

A major part of modern weather forecasting is the severe weather alerts and advisories which the national weather services issue in the case that severe or hazardous weather is expected. This is done to protect life and property. Some of the most commonly known of severe weather advisories are the severe thunderstorm and tornado warning, as well as the severe thunderstorm and tornado watch. Other forms of these advisories include winter weather, high wind, flood, tropical cyclone, and fog. Severe weather advisories and alerts are broadcast through the media, including radio, using emergency systems as the Emergency Alert System which break into regular programming.

Low Temperature Forecast

The low temperature forecast for the current day is calculated using the lowest temperature found between 7pm that evening through 7am the following morning. So, in short, today's forecasted low is most likely tomorrow's low temperature.

Specialist Forecasting

There are a number of sectors with their own specific needs for weather forecasts and specialist services are provided to these users.

Air Traffic

Because the aviation industry is especially sensitive to the weather, accurate weather forecasting is essential. Fog or exceptionally low ceilings can prevent many aircraft from landing and taking off. Turbulence and icing are also significant in-flight hazards. Thunderstorms are a problem for all aircraft because of severe turbulence due to their updrafts and outflow boundaries, icing due to the heavy precipitation, as well as large hail, strong winds, and lightning, all of which can cause severe damage to an aircraft in flight. Volcanic ash is also a significant problem for aviation, as aircraft can lose engine power within ash clouds. On a day-to-day basis airliners are routed to take advantage of the jet stream tailwind to improve fuel efficiency. Aircrews are briefed prior to takeoff on the conditions to expect en route and at their destination. Additionally, airports often change which runway is being used to take advantage of a headwind. This reduces the distance required for takeoff, and eliminates potential crosswinds.

Ash cloud from the 2008 eruption of Chaitén volcano stretching across Patagonia from the Pacific to the Atlantic Ocean

Marine

Commercial and recreational use of waterways can be limited significantly by wind direction and speed, wave periodicity and heights, tides, and precipitation. These factors can each influence the safety of marine transit. Consequently, a variety of codes have been established to efficiently transmit detailed marine weather forecasts to vessel pilots via radio, for example the MAFOR (marine forecast). Typical weather forecasts can be received at sea through the use of RTTY, Navtex and Radiofax.

Agriculture

Farmers rely on weather forecasts to decide what work to do on any particular day. For example, drying hay is only feasible in dry weather. Prolonged periods of dryness can ruin cotton, wheat, and corn crops. While corn crops can be ruined by drought, their dried remains can be used as a cattle feed substitute in the form of silage. Frosts and freezes play havoc with crops both during the spring and fall. For example, peach trees in full bloom can have their potential peach crop decimated by a spring freeze. Orange groves can suffer significant damage during frosts and freezes, regardless of their timing.

Forestry

Weather forecasting of wind, precipitations and humidity is essential for preventing and controlling wildfires. Different indices, like the *Forest fire weather index* and the *Haines Index*, have been developed to predict the areas more at risk to experience fire from natural or human causes. Conditions for the development of harmful insects can be predicted by forecasting the evolution of weather, too.

Utility Companies

An air handling unit is used for the heating and cooling of air in a central location (click on image for legend).

Electricity and gas companies rely on weather forecasts to anticipate demand which can be strongly affected by the weather. They use the quantity termed the degree day to determine how strong of a use there will be for heating (heating degree day) or cooling (cooling degree day). These quantities are based on a daily average temperature of 65 °F (18 °C). Cooler temperatures force heating degree days (one per degree Fahrenheit), while warmer temperatures force cooling degree days. In winter, severe cold weather can cause a surge in demand as people turn up their heating. Similarly, in summer a surge in demand can be linked with the increased use of air conditioning systems in hot weather. By anticipating a surge in demand, utility companies can purchase additional supplies of power or natural gas before the price increases, or in some circumstances, supplies are restricted through the use of brownouts and blackouts.

Other Commercial Companies

Increasingly, private companies pay for weather forecasts tailored to their needs so that they can increase their profits or avoid large losses. For example, supermarket chains may change the stocks on their shelves in anticipation of different consumer spending habits in different weather conditions. Weather forecasts can be used to invest in the commodity market, such as futures in oranges, corn, soybeans, and oil.

Military Applications

United Kingdom Armed Forces

The UK Royal Navy, working with the UK Met Office, has its own specialist branch of weather observers and forecasters, as part of the Hydrographic and Meteorological (HM) specialisation, who monitor and forecast operational conditions across the globe, to provide accurate and timely weather and oceanographic information to submarines, ships and Fleet Air Arm aircraft.

Royal Air Force

A mobile unit in the RAF, working with the UK Met Office, forecasts the weather for regions in which British, allied servicemen and women are deployed. A group based at Camp Bastion provides forecasts for the British armed forces in Afghanistan.

United States Armed Forces

US Navy

Emblem of JTWC Joint Typhoon Warning Center

Similar to the private sector, military weather forecasters present weather conditions to the war fighter community. Military weather forecasters provide pre-flight and in-flight weather briefs to pilots and provide real time resource protection services for military installations. Naval forecasters cover the waters and ship weather forecasts. The United States Navy provides a special service to both themselves and the rest of the federal government by issuing forecasts for tropical cyclones across the Pacific and Indian Oceans through their Joint Typhoon Warning Center.

US Air Force

Within the United States, Air Force Weather provides weather forecasting for the Air Force and the Army. Air Force forecasters cover air operations in both wartime and peacetime operations and provide Army support; United States Coast Guard marine science technicians provide ship forecasts for ice breakers and other various operations within their realm; and Marine forecasters provide support for ground- and air-based United States Marine Corps operations. All four military branches take their initial enlisted meteorology technical training at Keesler Air Force Base. Military and civilian forecasters actively cooperate in analyzing, creating and critiquing weather forecast products.

Weather Map

A weather map displays various meteorological features across a particular area at a particular point in time and has various symbols which all have specific meanings. Such maps have been in use since the mid-19th century and are used for research and weather forecasting purposes. Maps using isotherms show temperature gradients, which can help locate weather fronts. Isotach maps, analyzing lines of equal wind speed, on a constant pressure surface of 300 mb or 250 mb show where the jet stream is located. Use of constant pressure charts at the 700 and 500 hPa level can indicate tropical cyclone motion. Two-dimensional streamlines based on wind speeds at various levels show areas of convergence and divergence in the wind field, which are helpful in determining the location of features within the wind pattern. A popular type of surface weather map is the surface weather analysis, which plots isobars to depict areas of high pressure and low pressure. Cloud codes are translated into symbols and plotted on these maps along with other meteorological data that are included in synoptic reports sent by professionally trained observers.

History

Sir Francis Galton, the inventor of the weather map.

The use of weather charts in a modern sense began in the middle portion of the 19th century in order to devise a theory on storm systems. During the Crimean War a storm devastated the French fleet at Balaklava, and the French scientist Urbain Le Verrier was able to show that if a chronological map of the storm had been issued, the path it would take could have been predicted and avoided by the fleet.

In England, the scientist Francis Galton heard of this work, as well as the pioneering weather forecasts of Robert Fitzroy. After gathering information from weather stations across the country for the month of October 1861, he plotted the data on a map using his own system of symbols, thereby creating the world's first weather map. He used his map to prove that air circulated clockwise around areas of high pressure; he coined the term 'anticyclone' to describe the phenomenon. He was also instrumental in publishing the first weather map in a newspaper, for which he modified the pantograph (an instrument for copying drawings) to inscribe the map onto printing blocks. *The Times* began printing weather maps using these methods with data from the Meteorological Office.

The introduction of country-wide weather maps required the existence of national telegraph networks so that data from across the country could be gathered in real time and remain relevant for subsequent analysis. The first such use of the telegraph for gathering data on the weather was the *Manchester Examiner* newspaper in 1847:

> ...led us to inquire if the electric telegraph was yet extended far enough from Manchester to obtain information from the eastern counties...inquiries were made at the following places; and answers were returned, which we append...

It was also important for time to be standardized across time zones so that the information on the map should accurately represent the weather at a given time. A standardized time system was first used to coordinate the British railway network in 1847, with the inauguration of Greenwich Mean Time.

In the USA, The Smithsonian Institution developed its network of observers over much of the central and eastern United States between the 1840s and 1860s once Joseph Henry took the helm. The U.S. Army Signal Corps inherited this network between 1870 and 1874 by an act of Congress, and expanded it to the west coast soon afterwards. At first, not all the data on the map was used due to a lack of time standardization. The United States fully adopted time zones in 1905, when Detroit finally established standard time.

20th century

Light tables were important to the construction of surface weather analyses into the 1990s

The use of frontal zones on weather maps began in the 1910s in Norway. Polar front theory is attributed to Jacob Bjerknes, derived from a coastal network of observation sites in Norway during World War I. This theory proposed that the main inflow into a cyclone was concentrated along two lines of convergence, one ahead of the low and another trailing behind the low. The convergence line ahead of the low became known as either the steering line or the warm front. The trailing convergence zone was referred to as the squall line or cold front. Areas of clouds and rainfall appeared to be focused along these convergence zones. The concept of frontal zones led to the concept of air masses. The nature of the three-dimensional structure of the cyclone would wait for the development of the upper air network during the 1940s. Since the leading edge of air mass changes bore resemblance to the military fronts of World War I, the term "front" came into use to represent these lines. The United States began to formally analyze fronts on surface analyses in late 1942, when the WBAN Analysis Center opened in downtown Washington, D.C.

In addition to surface weather maps, weather agencies began to generate constant pressure charts. In 1948, the United States began the Daily Weather Map series, which at first analyzed the 700 hPa level, which is around 3,000 metres (9,800 ft) above sea level. By May 14, 1954, the 500 hPa surface was being analyzed, which is about 5,520 metres (18,110 ft) above sea level. The effort to automate map plotting began in the United States in 1969, with the process complete in the 1970s. Hong Kong completed their process of automated surface plotting by 1987.

By 1999, computer systems and software had finally become sophisticated enough to allow for the ability to underlay on the same workstation satellite imagery, radar imagery, and model-derived fields such as atmospheric thickness and frontogenesis in combination with surface observations to make for the best possible surface analysis. In the United States, this development was achieved when Intergraph workstations were replaced by n-AWIPS workstations. By 2001, the various surface analyses done within the National Weather Service were combined into the Unified Surface Analysis, which is issued every six hours and combines the analyses of four different centers. Recent advances in both the fields of meteorology and geographic information systems have made it possible to devise finely tailored products that take us from the traditional weather map into an entirely new realm. Weather information can quickly be matched to relevant geographical detail. For instance, icing conditions can be mapped onto the road network. This will likely continue to lead to changes in the way surface analyses are created and displayed over the next several years.

Plotting of Data

Present weather symbols used on weather maps

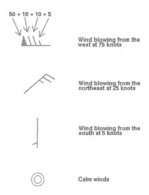

Wind barb interpretation

A station model is a symbolic illustration showing the weather occurring at a given reporting station. Meteorologists created the station model to plot a number of weather elements in a small space on weather maps. Maps filled with dense station-model plots can be difficult to read, but they allow meteorologists, pilots, and mariners to see important weather patterns. A computer draws a station model for each observation location. The station model is primarily used on surface-weather maps, but can also be used to show the weather aloft. A completed station-model map allows users to analyze patterns in air pressure, temperature, wind, cloud cover, and precipitation.

Station model plots use an internationally accepted coding convention that has changed little since

August 1, 1941. Elements in the plot show the key weather elements, including temperature, dewpoint, wind, cloud cover, air pressure, pressure tendency, and precipitation. Winds have a standard notation when plotted on weather maps. More than a century ago, winds were plotted as arrows, with feathers on just one side depicting five knots of wind, while feathers on both sides depicted 10 knots (19 km/h) of wind. The notation changed to that of half of an arrow, with half of a wind barb indicating five knots, a full barb ten knots, and a pennant flag fifty knots.

Because of the structure of the SYNOP code, a maximum of three cloud symbols can be plotted for each reporting station that appears on the weather map. All cloud types are coded and transmitted by trained observers then plotted on maps as low, middle, or high-étage using special symbols for each major cloud type. Any cloud type with significant vertical extent that can occupy more than one étage is coded as low (cumulus and cumulonimbus) or middle (nimbostratus) depending on the altitude level or étage where it normally initially forms aside from any vertical growth that takes place. The symbol used on the map for each of these étages at a particular observation time is for the genus, species, variety, mutation, or cloud motion that is considered most important according to criteria set out by the World Meteorological Organization (WMO). If these elements for any étage at the time of observation are deemed to be of equal importance, then the type which is predominant in amount is coded by the observer and plotted on the weather map using the appropriate symbol. Special weather maps in aviation show areas of icing and turbulence.

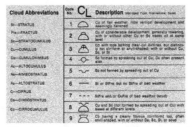

Low étage (Sc,St) and upward-growing vertical (Cu, Cb)

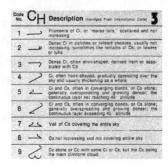

Middle étage (Ac,As) and downward-growing vertical (Ns)

High étage (Ci,Cc,Cs)

Types

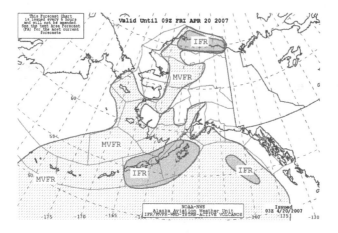

Alaskan aviation weather map

Aviation Maps

Aviation interests have their own set of weather maps. One type of map shows where VFR (visual flight rules) are in effect and where IFR (instrument flight rules) are in effect. Weather depiction plots show ceiling height (level where at least half the sky is covered with clouds) in hundreds of feet, present weather, and cloud cover. Icing maps depict areas where icing can be a hazard for flying. Aviation-related maps also show areas of turbulence.

Constant Pressure Charts

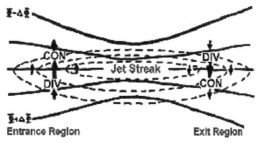

An upper level jet streak. DIV areas are regions of divergence aloft, which usually leads to surface convergence and cyclogenesis.

Constant pressure charts normally contain plotted values of temperature, humidity, wind, and the vertical height above sea level of the pressure surface. They have a variety of uses. In the mountainous terrain of the western United States and Mexican Plateau, the 850 hPa pressure surface can be a more realistic depiction of the weather pattern than a standard surface analysis. Using the 850 and 700 hPa pressure surfaces, one can determine when and where warm advection (coincident with upward vertical motion) and cold advection (coincident with downward vertical motion) is occurring within the lower portions of the troposphere. Areas with small dewpoint depressions and are below freezing indicate the presence of icing conditions for aircraft. The 500 hPa pressure surface can be used as a rough guide for the motion of many tropical cyclones. Shallower tropical cyclones, which have experienced vertical wind shear, tend to be steered by winds at the 700 hPa level.

Use of the 300 and 200 hPa constant pressure charts can indicate the strength of systems in the lower troposphere, as stronger systems near the Earth's surface are reflected as stronger features at these levels of the atmosphere. Isotachs are drawn at these levels, which a lines of equal wind speed. They are helpful in finding maxima and minima in the wind pattern. Minima in the wind pattern aloft are favorable for tropical cyclogenesis. Maxima in the wind pattern at various levels of the atmosphere show locations of jet streams. Areas colder than −40 °C (−40 °F) indicate a lack of significant icing, as long as there is no active thunderstorm activity.

Surface Weather Analysis

A surface weather analysis is a type of weather map that depicts positions for high and low-pressure areas, as well as various types of synoptic scale systems such as frontal zones. Isotherms can be drawn on these maps, which are lines of equal temperature. Isotherms are drawn normally as solid lines at a preferred temperature interval. They show temperature gradients, which can be useful in finding fronts, which are on the warm side of large temperature gradients. By plotting the freezing line, isotherms can be useful in determination of precipitation type. Mesoscale boundaries such as tropical cyclones, outflow boundaries and squall lines also are analyzed on surface weather analyses.

Isobaric analysis is performed on these maps, which involves the construction of lines of equal mean sea level pressure. The innermost closed lines indicate the positions of relative maxima and minima in the pressure field. The minima are called low-pressure areas while the maxima are called high-pressure areas. Highs are often shown as H's whereas lows are shown as L's. Elongated areas of low pressure, or troughs, are sometimes plotted as thick, brown dashed lines down the trough axis. Isobars are commonly used to place surface boundaries from the horse latitudes poleward, while streamline analyses are used in the tropics. A streamline analysis is a series of arrows oriented parallel to wind, showing wind motion within a certain geographic area. C's depict cyclonic flow or likely areas of low pressure, while A's depict anticyclonic flow or likely positions of high-pressure areas. An area of confluent streamlines shows the location of shearlines within the tropics and subtropics.

Surface Weather Analysis

Surface weather analysis is a special type of weather map that provides a view of weather elements over a geographical area at a specified time based on information from ground-based weather stations.

Weather maps are created by plotting or tracing the values of relevant quantities such as sea level pressure, temperature, and cloud cover onto a geographical map to help find synoptic scale features such as weather fronts.

The first weather maps in the 19th century were drawn well after the fact to help devise a theory on storm systems. After the advent of the telegraph, simultaneous surface weather observations became possible for the first time, and beginning in the late 1840s, the Smithsonian Institution became the first organization to draw real-time surface analyses. Use of surface analyses began first in the United States, spreading worldwide during the 1870s. Use of the Norwegian cyclone model for frontal analysis began in the late 1910s across Europe, with its use finally spreading to the United States during World War II.

Surface weather analyses have special symbols that show frontal systems, cloud cover, precipitation, or other important information. For example, an *H* may represent high pressure, implying good and fair weather. An *L*, on the other hand, may represent low pressure, which frequently accompanies precipitation. Various symbols are used not just for frontal zones and other surface boundaries on weather maps, but also to depict the present weather at various locations on the weather map. Areas of precipitation help determine the frontal type and location.

History of Surface Analysis

The use of weather charts in a modern sense began in the middle portion of the 19th century in order to devise a theory on storm systems. The development of a telegraph network by 1845 made it possible to gather weather information from multiple distant locations quickly enough to preserve its value for real-time applications. The Smithsonian Institution developed its network of observers over much of the central and eastern United States between the 1840s and 1860s once Joseph Henry took the helm. The U.S. Army Signal Corps inherited this network between 1870 and 1874 by an act of Congress, and expanded it to the west coast soon afterwards.

Surface analysis of Great Blizzard of 1888 on March 12, 1888 at 10 pm

At first, all the data on the map was not taken from these analyses because of a lack of time standardization. The first attempts at time standardization took hold in Great Britain by 1855. The entire United States did not finally come under the influence of time zones until 1905, when Detroit finally established standard time. Other countries followed the lead of the United States in taking simultaneous weather observations, starting in 1873. Other countries then began preparing surface analyses. The use of frontal zones on weather maps did not appear until the introduction of the Norwegian cyclone model in the late 1910s, despite Loomis' earlier attempt at a similar notion in 1841. Since the leading edge of air mass changes bore resemblance to the military fronts of World War I, the term "front" came into use to represent these lines.

Despite the introduction of the Norwegian cyclone model just after World War I, the United States did not formally analyze fronts on surface analyses until late 1942, when the WBAN Analysis Center opened in downtown Washington, D.C.. The effort to automate map plotting began in the United States in 1969, with the process complete in the 1970s. Hong Kong completed their process of automated surface plotting by 1987. By 1999, computer systems and software had finally become sophisticated enough to allow for the ability to underlay on the same workstation satellite imagery, radar imagery, and model-derived fields such as atmospheric thickness and frontogenesis in combination with surface observations to make for the best possible surface analysis. In the United States, this development was achieved when Intergraph workstations were replaced by n-AWIPS workstations. By 2001, the various surface analyses done within the National Weather Service were combined into the Unified Surface Analysis, which is issued every six hours and combines the analyses of four different centers. Recent advances in both the fields of meteorology and geographic information systems have made it possible to devise finely tailored products that take us from the traditional weather map into an entirely new realm. Weather information can quickly be matched to relevant geographical detail. For instance, icing conditions can be mapped onto the road network. This will likely continue to lead to changes in the way surface analyses are created and displayed over the next several years. The pressureNET project is an ongoing attempt to gather surface pressure data using smartphones.

Station Model Used on Weather Maps

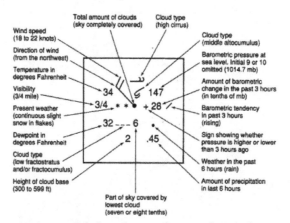

Station model plotted on surface weather analyses

When analyzing a weather map, a station model is plotted at each point of observation. Within the station model, the temperature, dewpoint, wind speed and direction, atmospheric pressure, pressure tendency, and ongoing weather are plotted. The circle in the middle represents cloud cover. If completely filled in, it is overcast. If conditions are completely clear, the circle is empty. If conditions are partly cloudy, the circle is partially filled in. Outside the United States, temperature and dewpoint are plotted in degrees Celsius. Each full flag on the Wind Barb represents 10 knots (19 km/h) of wind, each half flag represents 5 knots (9 km/h). When winds reach 50 knots (93 km/h), a filled in triangle is used for each 50 knots (93 km/h) of wind. In the United States, rainfall plotted in the corner of the station model are in English units, inches. The international standard rainfall measurement unit is the millimeter. Once a map has a field of station models plotted, the analyzing isobars (lines of equal pressure), isallobars (lines of equal pressure change), isotherms (lines of equal temperature), and isotachs (lines of equal wind speed) can be easily accomplished. The abstract present weather symbols used on surface weather analyses for obstructions to visibility, precipitation, and thunderstorms were devised to take up the least room possible on weather maps.

Synoptic Scale Features

A synoptic scale feature is one whose dimensions are large in scale, more than several hundred kilometers in length. Migratory pressure systems and frontal zones exist on this scale.

Pressure Centers

Centers of surface high- and low-pressure areas are found within closed isobars on a surface weather analysis where they are the absolute maxima and minima in the pressure field, and can tell a user in a glance what the general weather is in their vicinity. Weather maps in English-speaking countries will depict their highs as Hs and lows as Ls, while Spanish-speaking countries will depict their highs as As and lows as Bs.

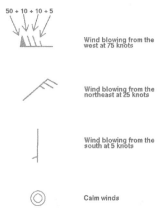

50 + 10 + 10 + 5

Wind blowing from the west at 75 knots

Wind blowing from the northeast at 25 knots

Wind blowing from the south at 5 knots

Calm winds

Wind barb interpretation

Low Pressure

Low-pressure systems, also known as cyclones, are located in minima in the pressure field. Rotation is inward and counterclockwise in the northern hemisphere as opposed to inward and clockwise in the southern hemisphere due to the Coriolis force. Weather is normally unsettled in the vicinity of a cyclone, with increased cloudiness, increased winds, increased temperatures, and upward motion in the atmosphere, which leads to an increased chance of precipitation. Polar lows can form over relatively mild ocean waters when cold air sweeps in from the ice cap, leading to upward motion and convection, usually in the form of snow. Tropical cyclones and winter storms are intense varieties of low pressure. Over land, thermal lows are indicative of hot weather during the summer.

High Pressure

High-pressure systems, also known as anticyclones, rotate outward and clockwise in the northern hemisphere as opposed to outward and counterclockwise in the southern hemisphere. Under surface highs, sinking motion leads to skies that are clearer, winds that are lighter, and there is a reduced chance of precipitation. There is normally a greater range between high and low temperature due to the drier air mass present. If high pressure persists, air pollution will build up due to pollutants trapped near the surface caused by the subsiding motion associated with the high.

Fronts

Fronts in meteorology are the leading edges of air masses with different density (e.g., air temperature and/or humidity). When a front passes over an area, it is marked by changes in temperature, moisture, wind speed and direction, atmospheric pressure, and often a change in the precipitation pattern. Cold fronts are closely associated with low pressure systems, normally lying at the leading edge of high-pressure systems and, in the case of the polar front, at approximately the equatorward edge of the high-level polar jet. Fronts are guided by winds aloft, but they normally move at lesser speeds. In the northern hemisphere, they usually travel from west to east (though they can move in a north-south direction as well). Movement is due to the pressure gradient force (horizontal differences in atmospheric pressure) and the Coriolis effect, caused by the Earth spinning about its axis. Frontal zones can be contorted by geographic features like mountains and large bodies of water.

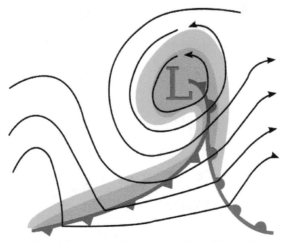

Occluded cyclone example. The triple point is the intersection of the cold, warm, and occluded fronts.

Cold Front

A cold front's location is at the leading edge of the temperature drop-off, which in an isotherm analysis shows up as the leading edge of the isotherm gradient, and it normally lies within a sharp surface trough. Cold fronts can move up to twice as fast as warm fronts and produce sharper changes in weather, since cold air is denser than warm air and rapidly replaces the warm air preceding the boundary. Cold fronts are typically accompanied by a narrow band of showers and thunderstorms. On weather maps, the surface position of the cold front is marked with the symbol of a blue line of triangles/spikes (pips) pointing in the direction of travel, and it is placed at the leading edge of the cooler air mass.

Warm Front

Warm fronts are at the trailing edge of the temperature increase, which is located on the equatorward edge of the gradient in isotherms, and lie within broader troughs of low pressure than cold fronts. Warm fronts move more slowly than the cold front that usually follows because cold air is denser, and harder to displace from the Earth's surface. This also forces temperature differences across warm fronts to be broader in scale. Clouds ahead of the warm front are mostly stratiform and rainfall gradually increases as the front approaches. Fog can also occur preceding a warm frontal passage. Clearing and warming is usually rapid after frontal passage. If the warm air mass is unstable, thunderstorms may be embedded among the stratiform clouds ahead of the front, and after frontal passage, thundershowers may continue. On weather maps, the surface location of a warm front is marked with a red line of half circles pointing in the direction of travel.

Illustration clouds overriding a warm front

Occluded Front

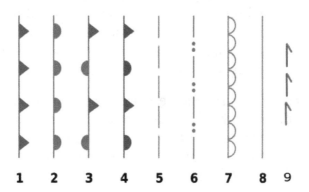

A guide to the symbols for weather fronts that may be found on a weather map:

1. cold front

2. warm front

3. stationary front

4. occluded front

5. surface trough

6. squall line

7. dry line

8. tropical wave

9. Trowal

An occluded front is formed during the process of cyclogenesis when a cold front overtakes a warm front. The cold and warm fronts curve naturally poleward into the point of occlusion, which is also known as the triple point in meteorology. It lies within a sharp trough, but the air mass behind the boundary can be either warm or cold. In a cold occlusion, the air mass overtaking the warm front is cooler than the cool air ahead of the warm front, and plows under both air masses. In a warm occlusion, the air mass overtaking the warm front is not as cool as the cold air ahead of the warm front, and rides over the colder air mass while lifting the warm air. A wide variety of weather can be found along an occluded front, with thunderstorms possible, but usually their passage is associated with a drying of the air mass. Occluded fronts are indicated on a weather map by a purple line with alternating half-circles and triangles pointing in direction of travel. Occluded fronts usually form around mature low pressure areas.

The trowal is the projection on the Earth's surface of a tongue of warm air aloft, such as may be formed during the occlusion process of a depression.

Stationary Fronts and Shearlines

A stationary front is a non-moving boundary between two different air masses, neither of which is strong enough to replace the other. They tend to remain in the same area for long periods of time, usually moving in waves. There is normally a broad temperature gradient behind the boundary with more widely spaced isotherm packing. A wide variety of weather can be found along a stationary front, but usually clouds and prolonged precipitation are found there. Stationary fronts will either dissipate after several days or devolve into shear lines, but can change into a cold or warm front if conditions aloft change. Stationary fronts are marked on weather maps with alternating red half-circles and blue spikes pointing in opposite directions, indicating no significant movement.

When stationary fronts become smaller in scale, degenerating to a narrow zone where wind direction changes over a short distance, they become known as shear lines. If the shear line becomes active with thunderstorms, it may support formation of a tropical storm or a regeneration of the feature back into a stationary front. A shear line is depicted as a line of red dots and dashes.

Mesoscale Features

Mesoscale features are smaller than synoptic scale systems like fronts, but larger than storm-scale systems like thunderstorms. Horizontal dimensions generally range from over ten kilometres to several hundred kilometres.

Dry Line

The dry line is the boundary between dry and moist air masses east of mountain ranges with similar orientation to the Rockies, depicted at the leading edge of the dew point, or moisture, gradient.

Near the surface, warm moist air that is denser than dry air of greater temperature wedges under the drier air like a cold front. When the warm moist air wedged under the drier mass heats up it becomes less dense than the drier air above and it begins to rise and sometimes forms thunderstorms. At higher altitudes, the warm moist air is less dense than the cooler, drier air and the boundary slope reverses. In the vicinity of the reversal aloft, severe weather is possible, especially when a triple point is formed with a cold front.

During daylight hours, drier air from aloft drifts down to the surface, causing an apparent movement of the dryline eastward. At night, the boundary reverts to the west as there is no longer any sunshine to help mix the lower atmosphere. If enough moisture converges upon the dryline, it can be the focus of afternoon and evening thunderstorms. A dry line is depicted on United States surface analyses as a brown line with scallops, or bumps, facing into the moist sector. Dry lines are one of the few surface fronts where the special shapes along the drawn boundary do not necessarily reflect the boundary's direction of motion.

Outflow Boundaries and Squall Lines

Organized areas of thunderstorm activity not only reinforce pre-existing frontal zones, but they can outrun cold fronts. This outrunning occurs in a pattern where the upper level jet splits into two streams. The resultant mesoscale convective system (MCS) forms at the point of the upper level split in the wind pattern in the area of best low-level inflow. The convection then moves east and equatorward into the warm sector, parallel to low-level thickness lines. When the convection is strong and linear or curved, the MCS is called a squall line, with the feature placed at the leading edge of the significant wind shift and pressure rise. Even weaker and less organized areas of thunderstorms will lead to locally cooler air and higher pressures, and outflow boundaries exist ahead of this type of activity, "SQLN" or "SQUALL LINE", while outflow boundaries are depicted as troughs with a label of "OUTFLOW BOUNDARY" or "OUTFLOW BNDRY".

A shelf cloud such as this one can be a sign that a squall is imminent

Sea and Land Breeze Fronts

Sea breeze fronts occur mainly on sunny days when the landmass warms up above the water temperature. Similar boundaries from downwind on lakes and rivers during the day, as well as offshore landmasses at night. Since the specific heat of water is so high, there is little diurnal change in bodies of water, even on the sunniest days. The water temperature varies less than 1 °C (1 to 2 °F). By contrast, the land, with a lower specific heat, can vary several degrees in a matter of hours.

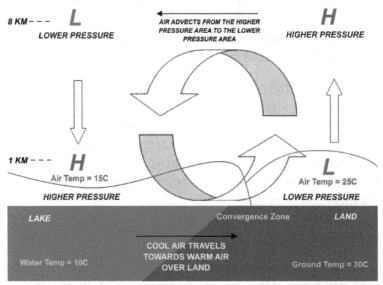

Idealized circulation pattern associated with a sea breeze

During the afternoon, air pressure decreases over the land as temperature rises. The relatively cooler air over the sea rushes in to fill the gap. The result is a relatively cool onshore wind. This process usually reverses at night where the water temperature is higher relative to the landmass, leading to an offshore land breeze. However, if water temperatures are colder than the land at night, the sea breeze may continue, only somewhat abated. This is typically the case along the California coast, for example.

If enough moisture exists, thunderstorms can form along sea breeze fronts that then can send out outflow boundaries. This causes chaotic wind/pressure regimes if the steering flow is light. Like all other surface features, sea breeze fronts lie inside troughs of low pressure.

References

- Fahd, Toufic. : 842 Missing or empty |title= (help); |contribution= ignored (help), in Rashed, Roshdi; Morelon, Régis (1996). Encyclopedia of the History of Arabic Science. 3. Routledge. pp. 813–852. ISBN 0-415-12410-7

- Ronalds, B.F. (2016). Sir Francis Ronalds: Father of the Electric Telegraph. London: Imperial College Press. ISBN 978-1-78326-917-4.

- Sunderam, V. S.; van Albada, G. Dick; Peter, M. A.; Sloot, J. J. Dongarra (2005). Computational Science – ICCS 2005: 5th International Conference, Atlanta, GA, USA, May 22–25, 2005, Proceedings, Part 1. Springer. p. 132. ISBN 978-3-540-26032-5. Retrieved 2011-01-02.

- Chan, Johnny C. L. & Jeffrey D. Kepert (2010). Global Perspectives on Tropical Cyclones: From Science to Mitigation. World Scientific. pp. 295–296. ISBN 978-981-4293-47-1. Retrieved 2011-02-24

- Holton, James R. (2004). An introduction to dynamic meteorology, Volume 1. Academic Press. p. 480. ISBN 978-0-12-354015-7. Retrieved 2011-02-24.

- Strikwerda, John C. (2004). Finite difference schemes and partial differential equations. SIAM. pp. 165–170. ISBN 978-0-89871-567-5. Retrieved 2010-12-31.

- John D. Cox (2002). Stormwatchers: The Turbulent History of Weather Prediction From Franklin's Kite to El

Nino. John Wiley & Sons, Inc. pp. 53–56. ISBN 0-471-38108-X.

- World Meteorological Organization, ed. (1975). Étages, International Cloud Atlas (PDF). I. pp. 15–16. ISBN 92-63-10407-7. Retrieved 26 August 2014.

- Terry T. Lankford (1999). Aircraft icing: a pilot's guide. McGraw-Hill Professional. pp. 129–134. ISBN 0-07-134139-0. Retrieved 2010-02-06.

- Weather Prediction Center (2007-03-01). "A Brief History of the Weather Prediction Center". National Oceanic and Atmospheric Administration. Retrieved 2014-07-01.

- Helen Czerski. "Orbit: Earth's Extraordinary Journey: 150 years since the first UK weather "forecast"". BBC. Retrieved November 5, 2013.

- Met Office (2012). "National Meteorological Library and Fact Sheet 8 -- The Shipping Forecast" (PDF). 1. pp. 3–5. Retrieved 2013-04-10.

- Allaby, Michael (2009). Atmosphere: A Scientific History of Air, Weather, and Climate. Infobase Publishing. Retrieved 2013-12-07.

- National Oceanic and Atmospheric Administration (2007-05-30). "An Expanding Presence". United States Department of Commerce. Retrieved 2010-01-31.

- Hong Kong Observatory (2009-09-03). "The Hong Kong Observatory Computer System and Its Applications". The Government of the Hong Kong Special Administrative Region. Retrieved 2010-02-06.

- National Weather Service Forecast Office Honolulu, Hawaii (2010-02-07). "Pacific Streamline Analysis". Pacific Region Headquarters. Retrieved 2010-02-07.

Evolution of Meteorology

This chapter chronicles the biggest scientific and technological advancements in the field of meteorology. It includes topics like observational meteorology, weather forecasting, climatology, atmospheric chemistry and atmospheric physics. It provides the reader with a timeline of meteorology spanning the ages till the present day.

Timeline of Meteorology

The timeline of meteorology contains events of scientific and technological advancements in the area of atmospheric sciences. The most notable advancements in observational meteorology, weather forecasting, climatology, atmospheric chemistry, and atmospheric physics are listed chronologically. Some historical weather events are included that mark time periods where advancements were made, or even that sparked policy change

Antiquity

- 3000 BC – Meteorology in India can be traced back to around 3000 BC, with writings such as the Upanishadas, containing discussions about the processes of cloud formation and rain and the seasonal cycles caused by the movement of earth round the sun.

- 600 BC – Thales may qualify as the first Greek meteorologist. He described the water cycle in a fairly accurate way. He also issued the first seasonal crop forecast.

- 400 BC – There is some evidence that Democritus predicted changes in the weather, and that he used this ability to convince people that he could predict other future events.

- 400 BC – Hippocrates writes a treatise called *Airs, Waters and Places*, the earliest known work to include a discussion of weather. More generally, he wrote about common diseases that occur in particular locations, seasons, winds and air.

- 350 BC – Aristotle writes *Meteorology*.

- Although the term meteorology is used today to describe a subdiscipline of the atmospheric sciences, Aristotle's work is more general. The work touches upon much of what is known as the earth sciences. In his own words:

  ...all the affections we may call common to air and water, and the kinds and parts of the earth and the affections of its parts.

  One of the most impressive achievements in *Meteorology* is his description of what is now known as the hydrologic cycle:

Aristotle

Now the sun, moving as it does, sets up processes of change and becoming and decay, and by its agency the finest and sweetest water is every day carried up and is dissolved into vapour and rises to the upper region, where it is condensed again by the cold and so returns to the earth.

- 250 BC – Archimedes studies the concepts of buoyancy and the hydrostatic principle. Positive buoyancy is necessary for the formation of convective clouds (cumulus, cumulus congestus and cumulonimbus).

- 25 AD – Pomponius Mela, a geographer for the Roman empire, formalizes the climatic zone system.

- c. 80 AD – In his *Lunheng* the Han Dynasty Chinese philosopher Wang Chong (27–97 AD) dispels the Chinese myth of rain coming from the heavens, and states that rain is evaporated from water on the earth into the air and forms clouds, stating that clouds condense into rain and also form dew, and says when the clothes of people in high mountains are moistened, this is because of the air-suspended rain water. However, Wang Chong supports his theory by quoting a similar one of Gongyang Gao's, the latter's commentary on the *Spring and Autumn Annals*, the Gongyang Zhuan, compiled in the 2nd century BC, showing that the Chinese conception of rain evaporating and rising to form clouds goes back much farther than Wang Chong. Wang Chong wrote:

As to this coming of rain from the mountains, some hold that the clouds carry the rain with them, dispersing as it is precipitated (and they are right). Clouds and rain are really the same thing. Water evaporating upwards becomes clouds, which condense into rain, or still further into dew.

Middle Ages

- 500 AD – In around 500 AD, the Indian astronomer, mathematician, and astrologer: Varāhamihira published his work Brihat-Samhita's, which provides clear evidence that a deep knowledge of atmospheric processes existed in the Indian region.

- 7th century – The poet Kalidasa in his epic Meghaduta, mentions the date of onset of the south-west Monsoon over central India and traces the path of the monsoon clouds.

- 7th century – St. Isidore of Seville,in his work *De Rerum Natura*, writes about astronomy, cosmology and meteorology. In the chapter dedicated to Meteorology, he discusses the thunder, clouds, rainbows and wind.

- 9th century – Al-Kindi (Alkindus), an Arab naturalist, writes a treatise on meteorology entitled *Risala fi l-Illa al-Failali l-Madd wa l-Fazr* (*Treatise on the Efficient Cause of the Flow and Ebb*), in which he presents an argument on tides which "depends on the changes which take place in bodies owing to the rise and fall of temperature."

- 9th century – Al-Dinawari, a Kurdish naturalist, writes the *Kitab al-Nabat* (*Book of Plants*), in which he deals with the application of meteorology to agriculture during the Muslim Agricultural Revolution. He describes the meteorological character of the sky, the planets and constellations, the Sun and Moon, the lunar phases indicating seasons and rain, the *anwa* (heavenly bodies of rain), and atmospheric phenomena such as winds, thunder, lightning, snow, floods, valleys, rivers, lakes, wells and other sources of water.

- 10th century – Ibn Wahshiyya's *Nabatean Agriculture* discusses the weather forecasting of atmospheric changes and signs from the planetary astral alterations; signs of rain based on observation of the lunar phases, nature of thunder and lightning, direction of sunrise, behaviour of certain plants and animals, and weather forecasts based on the movement of winds; pollenized air and winds; and formation of winds and vapours.

- 1021 – Ibn al-Haytham (Alhazen) writes on the atmospheric refraction of light, the cause of morning and evening twilight. He endeavored by use of hyperbola and geometric optics to chart and formulate basic laws on atmospheric refraction. He provides the first correct definition of the twilight, discusses atmospheric refraction, shows that the twilight is due to atmospheric refraction and only begins when the Sun is 19 degrees below the horizon, and uses a complex geometric demonstration to measure the height of the Earth's atmosphere as 52,000 *passuum* (49 miles), which is very close to the modern measurement of 50 miles.

- 1020s – Ibn al-Haytham publishes his *Risala fi l-Daw'* (*Treatise on Light*) as a supplement to his *Book of Optics*. He discusses the meteorology of the rainbow, the density of the atmosphere, and various celestial phenomena, including the eclipse, twilight and moonlight.

- 1027 – Avicenna publishes *The Book of Healing*, in which Part 2, Section 5, contains his essay on mineralogy and meteorology in six chapters: formation of mountains; the advantages of mountains in the formation of clouds; sources of water; origin of earthquakes; formation of minerals; and the diversity of earth's terrain. He also describes the structure of a meteor, and his theory on the formation of metals combined Jābir ibn Hayyān's sulfur–mercury theory from Islamic alchemy (although he was critical of alchemy) with the mineralogical theories of Aristotle and Theophrastus. His scientific methodology of field observation was also original in the Earth sciences.

- Late 11th century – Abu 'Abd Allah Muhammad ibn Ma'udh, who lived in Al-Andalus, wrote a work on optics later translated into Latin as *Liber de crepisculis*, which was mistakenly attributed to Alhazen. This was a short work containing an estimation of the angle of depression of the sun at the beginning of the morning twilight and at the end of the evening twilight, and an attempt to calculate on the basis of this and other data the height of the atmospheric moisture responsible for the refraction of the sun's rays. Through his experiments, he obtained the accurate value of 18°, which comes close to the modern value.

- 1088 – In his *Dream Pool Essays*, the Chinese scientist Shen Kuo wrote vivid descriptions of tornadoes, that rainbows were formed by the shadow of the sun in rain, occurring when the sun would shine upon it, and the curious common phenomena of the effect of lightning that, when striking a house, would merely scorch the walls a bit but completely melt to liquid all metal objects inside.

- 1121 – Al-Khazini, a Muslim scientist of Byzantine Greek descent, publishes *The Book of the Balance of Wisdom*, the first study on the hydrostatic balance.

- 13th century-St. Albert the Great is the first to propose that each drop of falling rain had the form of a small sphere, and that this form meant that the rainbow was produced by light interacting with each raindrop.

- 1267 – Roger Bacon was the first to calculate the angular size of the rainbow. He stated that the rainbow summit can not appear higher than 42 degrees above the horizon.

- 1337 – William Merle, rector of Driby, starts recording his weather diary, the oldest existing in print. The endeavour ended 1344.

- Late 13th century – Theoderic of Freiburg and Kamāl al-Dīn al-Fārisī give the first accurate explanations of the primary rainbow, simultaneously but independently.Theoderic also gives the explanation for the secondary rainbow.

- 1441 – King Sejongs son, Prince Munjong, invented the first standardized rain gauge. These were sent throughout the Joseon Dynasty of Korea as an official tool to assess land taxes based upon a farmer's potential harvest.

- 1450 – Leone Battista Alberti developed a swinging-plate anemometer, and is known as the first *anemometer*.

- – Nicolas Cryfts, (Nicolas of Cusa), described the first hair hygrometer to measure humidity. The design was drawn by Leonardo da Vinci, referencing Cryfts design in *da Vinci's Codex Atlanticus*.

- 1488 – Johannes Lichtenberger publishes the first version of his *Prognosticatio* linking weather forecasting with astrology. The paradigm was only challenged centuries later.

- 1494 – During his second voyage Christopher Columbus experiences a tropical cyclone in the Atlantic Ocean, which leads to the first written European account of a hurricane.

- 1510 – Leonhard Reynmann, astronomer of Nuremberg, publishes "Wetterbüchlein Von warer erkanntnus des wetters", a collection of weather lore.

17th Century

Galileo.

- 1607 – Galileo Galilei constructs a thermoscope. Not only did this device measure temperature, but it represented a paradigm shift. Up to this point, heat and cold were believed to be qualities of Aristotle's elements (fire, water, air, and earth). *Note: There is some controversy about who actually built this first thermoscope. There is some evidence for this device being independently built at several different times.* This is the era of the first recorded meteorological observations. As there was no standard measurement, they were of little use until the work of Daniel Gabriel Fahrenheit and Anders Celsius in the 18th century.

Sir Francis Bacon

- 1611 – Johannes Kepler writes the first scientific treatise on snow crystals: "Strena Seu de Nive Sexangula (A New Year's Gift of Hexagonal Snow)".

- 1620 – Francis Bacon (philosopher) analyzes the scientific method in his philosophical work; Novum Organum.

- 1643 – Evangelista Torricelli invents the mercury barometer.

Blaise Pascal.

- 1648 – Blaise Pascal rediscovers that atmospheric pressure decreases with height, and deduces that there is a vacuum above the atmosphere.

- 1654 – Ferdinando II de Medici sponsors the first *weather observing* network, that consisted of meteorological stations in Florence, Cutigliano, Vallombrosa, Bologna, Parma, Milan, Innsbruck, Osnabrück, Paris and Warsaw. Collected data was centrally sent to Accademia del Cimento in Florence at regular time intervals.

- 1662 – Sir Christopher Wren invented the mechanical, self-emptying, tipping bucket rain gauge.

- 1667 – Robert Hooke builds another type of anemometer, called a pressure-plate anemometer.

- 1686 – Edmund Halley presents a systematic study of the trade winds and monsoons and identifies solar heating as the cause of atmospheric motions.

  – Edmund Halley establishes the relationship between barometric pressure and height above sea level.

18th Century

- 1716 – Edmund Halley suggests that aurorae are caused by "magnetic effluvia" moving along the Earth's magnetic field lines.

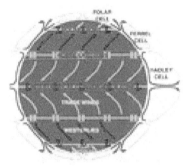

Global circulation as described by Hadley.

- 1724 – Gabriel Fahrenheit creates reliable scale for measuring temperature with a mercury-type thermometer.

- 1735 – The first *ideal* explanation of global circulation was the study of the Trade winds by George Hadley.

- 1738 – Daniel Bernoulli publishes *Hydrodynamics*, initiating the kinetic theory of gases. He gave a poorly detailed equation of state, but also the basic laws for the theory of gases.

- 1742 – Anders Celsius, a Swedish astronomer, proposed the Celsius temperature scale which led to the current Celsius scale.

- 1743 – Benjamin Franklin is prevented from seeing a lunar eclipse by a hurricane, he decides that cyclones move in a contrary manner to the winds at their periphery.

- 1761 – Joseph Black discovers that ice absorbs heat without changing its temperature when melting.

- 1772 – Black's student Daniel Rutherford discovers nitrogen, which he calls *phlogisticated air*, and together they explain the results in terms of the phlogiston theory.

- 1774 – Louis Cotte is put in charge of a "medico-meteorological" network of French veterinarians and country doctors to investigate the relationship between plague and weather. The project continued until 1794.

- - Royal Society begins twice daily observations compiled by Samuel Horsley testing for the influence of winds and of the moon on the barometer readings.

- 1777 – Antoine Lavoisier discovers oxygen and develops an explanation for combustion.

- 1780 – Charles Theodor charters the first international network of meteorological observers known as "Societas Meteorologica Palatina". The project collapses in 1795.

- 1780 - James Six invents the Six's thermometer, a thermometer that records minimum and maximum temperatures.

- 1783 – In Lavoisier's article "Reflexions sur le phlogistique", he deprecates the phlogiston theory and proposes a caloric theory of heat.

  – First hair hygrometer demonstrated. The inventor was Horace-Bénédict de Saussure.

19th Century

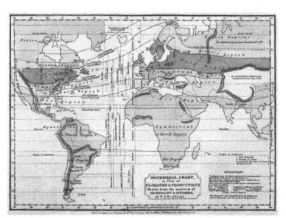

Isothermal chart of the world created 1823 by William Channing Woodbridge using the work of Alexander von Humboldt.

- 1800 – The Voltaic pile was the first modern electric battery, invented by Alessandro Volta, which led to later inventions like the telegraph.

- 1802–1803 – Luke Howard writes *On the Modification of Clouds* in which he assigns cloud types Latin names.

- 1804 – Sir John Leslie observes that a matte black surface radiates heat more effectively than a polished surface, suggesting the importance of black body radiation.

- 1806 – Francis Beaufort introduces his system for classifying wind speeds.

- 1808 – John Dalton defends caloric theory in *A New System of Chemistry* and describes how it combines with matter, especially gases; he proposes that the heat capacity of gases varies inversely with atomic weight.

- 1810 – Sir John Leslie freezes water to ice artificially.

- 1817 – Alexander von Humboldt publishes a global map of average temperature, the first global climate analysis.

- 1819 – Pierre Louis Dulong and Alexis Thérèse Petit give the Dulong-Petit law for the specific heat capacity of a crystal.

- 1820 – Heinrich Wilhelm Brandes publishes the first synoptic weather maps.

- – John Herapath develops some ideas in the kinetic theory of gases but mistakenly associates temperature with molecular momentum rather than kinetic energy; his work receives little attention other than from Joule.

- 1822 – Joseph Fourier formally introduces the use of dimensions for physical quantities in his *Theorie Analytique de la Chaleur*.

- 1824 – Sadi Carnot analyzes the efficiency of steam engines using caloric theory; he develops the notion of a reversible process and, in postulating that no such thing exists in nature, lays the foundation for the second law of thermodynamics.

- 1827 – Robert Brown discovers the Brownian motion of pollen and dye particles in water.

- 1832 – An electromagnetic telegraph was created by Baron Schilling.

- 1834 – Émile Clapeyron popularises Carnot's work through a graphical and analytic formulation.

- 1835 – Gaspard-Gustave Coriolis publishes theoretical discussions of machines with revolving parts and their efficiency, for example the efficiency of waterweels. At the end of the 19th century, meteorologists recognized that the way the Earth's rotation is taken into account in meteorology is analogous to what Coriolis discussed: an example of Coriolis Effect.

- 1836 – An American scientist, Dr. David Alter, invented the first known American electric telegraph in Elderton, Pennsylvania, one year before the much more popular Morse telegraph was invented.

- 1837 – Samuel Morse independently developed an electrical telegraph, an alternative design that was capable of transmitting over long distances using poor quality wire. His assistant, Alfred Vail, developed the Morse code signaling alphabet with Morse. The first electric telegram using this device was sent by Morse on May 24, 1844 from the U.S. Capitol in Washington, D.C. to the B&O Railroad "outer depot" in Baltimore and sent the message:

  What hath God wrought

- 1839 – The *first commercial* electrical telegraph was constructed by Sir William Fothergill Cooke and entered use on the Great Western Railway. Cooke and Wheatstone patented it in May 1837 as an alarm system.

- 1840 – Elias Loomis the first person known to attempt to devise a theory on frontal zones. The idea of fronts did not catch on until expanded upon by the Norwegians in the years following World War I.

- 1843 – John James Waterston fully expounds the kinetic theory of gases, but is ridiculed and ignored.

- – James Prescott Joule experimentally finds the mechanical equivalent of heat.

- 1844 – Lucien Vidi invented the aneroid, *from Greek meaning without liquid*, barometer.

- 1845 – Francis Ronalds invented the first successful camera for continuous recording of the variations in meteorological parameters over time

- 1845 – Francis Ronalds invented and named the storm clock, used to monitor rapid changes in meteorological parameters during extreme events

- 1846 – Cup anemometer invented by Dr. John Thomas Romney Robinson.

- 1847 – Francis Ronalds and William Radcliffe Birt described a stable kite to make observations at altitude using self-recording instruments

- 1847 – Hermann von Helmholtz publishes a definitive statement of the conservation of energy, the first law of thermodynamics.

- – The Manchester Examiner newspaper organises the first weather reports collected by electrical means.

- 1848 – William Thomson extends the concept of absolute zero from gases to all substances.

- 1849 – Smithsonian Institution begins to establish an observation network across the United States, with 150 observers via telegraph, under the leadership of Joseph Henry.

- – William John Macquorn Rankine calculates the correct relationship between saturated vapour pressure and temperature using his *hypothesis of molecular vortices*.

- 1850 – Rankine uses his *vortex* theory to establish accurate relationships between the temperature, pressure, and density of gases, and expressions for the latent heat of evaporation of a liquid; he accurately predicts the surprising fact that the apparent specific heat of saturated steam will be negative.

- – Rudolf Clausius gives the first clear joint statement of the first and second law of thermodynamics, abandoning the caloric theory, but preserving Carnot's principle.

- 1852 – Joule and Thomson demonstrate that a rapidly expanding gas cools, later named the Joule-Thomson effect.

- 1853 – The first International Meteorological Conference was held in Brussels at the initiative of Matthew Fontaine Maury, U.S. Navy, recommending standard observing times, methods of observation and logging format for weather reports from ships at sea.

- 1854 – The French astronomer Leverrier showed that a storm in the Black Sea could be followed across Europe and would have been predictable if the telegraph had been used. A service of storm forecasts was established a year later by the Paris Observatory.

– Rankine introduces his *thermodynamic function*, later identified as entropy.

- 1856 – William Ferrel publishes his essay on the winds and the currents of the oceans.

- 1859 – James Clerk Maxwell discovers the distribution law of molecular velocities.

- 1860 – Robert FitzRoy uses the new telegraph system to gather daily observations from across England and produces the first synoptic charts. He also coined the term "weather forecast" and his were the first ever daily weather forecasts to be published in this year.

  – After establishment in 1849, 500 U.S. telegraph stations are now making weather observations and submitting them back to the Smithsonian Institution. The observations are later interrupted by the American Civil War.

- 1865 – Josef Loschmidt applies Maxwell's theory to estimate the number-density of molecules in gases, given observed gas viscosities.

- – Manila Observatory founded in the Philippines.

- 1869 – Joseph Lockyer starts the scientific journal *Nature*.

- 1869 – The New York Meteorological Observatory opens, and begins to record wind, precipitation and temperature data.

- 1870 – The US Weather Bureau is founded. Data recorded in several Midwestern cities such as Chicago begins.

- 1870 – Benito Viñes becomes the head of the Meteorological Observatory at Belen in Havana, Cuba. He develops the first observing network in Cuba and creates some of the first hurricane-related forecasts.

- 1872 – The "Oficina Meteorológica Argentina" (today "Argentinean National Weather Service") is founded.

- 1872 – Ludwig Boltzmann states the Boltzmann equation for the temporal development of distribution functions in phase space, and publishes his H-theorem.

- 1873 – International Meteorological Organization formed in Vienna.

- – United States Army Signal Corp, forerunner of the National Weather Service, issues its first hurricane warning.

- 1875 – The India Meteorological Department is established, after a tropical cyclone struck Calcutta in 1864 and monsoon failures during 1866 and 1871.

- 1876 – Josiah Willard Gibbs publishes the first of two papers (the second appears in 1878) which discuss phase equilibria, statistical ensembles, the free energy as the driving force behind chemical reactions, and chemical thermodynamics in general.

- 1881 – Finnish Meteorological Central Office was formed from part of Magnetic Observatory of Helsinki University.

- 1890 – US Weather Bureau is created as a civilian operation under the U.S. Department of Agriculture.

- 1892 – William Henry Dines invented another kind of anemometer, called the pressure-tube (Dines) anemometer. His device measured the difference in pressure arising from wind blowing in a tube versus that blowing across the tube.

– The first mention of the term "El Niño" to refer to climate occurs when Captain Camilo Carrilo told the Geographical society congress in Lima that Peruvian sailors named the warm northerly current "El Niño" because it was most noticeable around Christmas.

- 1896 – IMO publishes the first International cloud atlas.

- – Svante Arrhenius proposes carbon dioxide as a key factor to explain the ice ages.

- 1898 – US Weather Bureau established a hurricane warning network at Kingston, Jamaica.

20th Century

- 1902 – Richard Assmann and Léon Teisserenc de Bort, two European scientists, independently discovered the stratosphere.

  - The Marconi Company issues the first routine weather forecast by means of radio to ships on sea. Weather reports from ships started 1905.

- 1903 – Max Margules publishes „Über die Energie der Stürme", an essay on the atmosphere as a three-dimensional thermodynamical machine.

- 1904 – Vilhelm Bjerknes presents the vision that forecasting the weather is feasible based on mathematical methods.

- 1905 – Australian Bureau of Meteorology established by a Meteorology Act to unify existing state meteorological services.

- 1919 – Norwegian cyclone model introduced for the first time in meteorological literature. Marks a revolution in the way the atmosphere is conceived and immediately starts leading to improved forecasts.

  - Sakuhei Fujiwhara is the first to note that hurricanes move with the larger scale flow, and later publishes a paper on the Fujiwhara effect in 1921.

- 1920 – Milutin Milanković proposes that long term climatic cycles may be due to changes in the eccentricity of the Earth's orbit and changes in the Earth's obliquity.

- 1922 – Lewis Fry Richardson organises the first numerical weather prediction experiment.

- 1923 – The oscillation effects of ENSO were first *erroneously* described by Sir Gilbert Thomas Walker from whom the Walker circulation takes its name; now an important aspect of the *Pacific ENSO* phenomenon.

- 1924 – Gilbert Walker first coined the term "Southern Oscillation".

- 1930, January 30 – Pavel Molchanov invents and launches the first radiosonde. Named "271120", it was released 13:44 Moscow Time in Pavlovsk, USSR from the Main Geophysical Observatory, reached a height of 7.8 kilometers measuring temperature there (−40.7 °C) and sent the first aerological message to the Leningrad Weather Bureau and Moscow Central Forecast Institute.

- 1935 – IMO decides on the 30 years normal period (1900–1930) to describe the climate.

- 1937 – The U.S. Army Air Forces Weather Service was established (redesignated in 1946 as AWS-Air Weather Service).

- 1938 – Guy Stewart Callendar first to propose global warming from carbon dioxide emissions.

- 1939 – Rossby waves were first identified in the atmosphere by Carl-Gustaf Arvid Rossby who explained their motion. Rossby waves are a subset of inertial waves.

- 1941 – Pulsed radar network is implemented in England during World War II. Generally during the war, operators started noticing echoes from weather elements such as rain and snow.

- 1943 – 10 years after flying into the Washington Hoover Airport on mainly instruments during the August 1933 Chesapeake-Potomac hurricane, J. B. Duckworth flies his airplane into a Gulf hurricane off the coast of Texas, proving to the military and meteorological community the utility of weather reconnaissance.

- 1944 – The Great Atlantic Hurricane is caught on radar near the Mid-Atlantic coast, the first such picture noted from the United States.

- 1947 – The Soviet Union launched its first Long Range Ballistic Rocket October 18, based on the German rocket A4 (V-2). The photographs demonstrated the immense potential of observing weather from space.

- 1948 – First correct tornado prediction by Robert C. Miller and E. J. Fawbush for tornado in Oklahoma.

  – Erik Palmén publishes his findings that hurricanes require surface water temperatures of at least 26°C (80°F) in order to form.

- 1950 – First successful numerical weather prediction experiment. Princeton University, group of Jule Gregory Charney on ENIAC.

  – Hurricanes begin to be named alphabetically with the radio alphabet.

  – WMO World Meteorological Organization replaces IMO under the auspice of the United Nations.

- 1953 – National Hurricane Center (NOAA) creates a system for naming hurricanes using alphabetical lists of women's names.

- 1954 – First routine real-time numerical weather forecasting. The Royal Swedish Air Force Weather Service.

  – A United States Navy rocket captures a picture of an inland tropical depression near the Texas/Mexico border, which leads to a surprise flood event in New Mexico. This convinces the government to set up a weather satellite program.

- 1955 – Norman Phillips at the Institute for Advanced Study in Princeton, New Jersey, runs first Atmospheric General Circulation Model.

– NSSP National Severe Storms Project and NHRP National Hurricane Research Projects established. The Miami office of the United States Weather Bureau is designated the main hurricane warning center for the Atlantic Basin.

- 1957–1958 – International Geophysical Year coordinated research efforts in eleven sciences, focused on polar areas during the solar maximum.

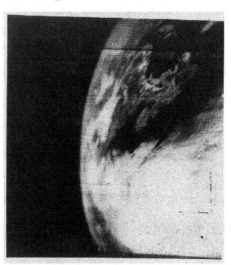

The first television image of Earth from space from the TIROS-1 weather satellite.

- 1959 – The first weather satellite, Vanguard 2, was launched on February 17. It was designed to measure cloud cover, but a poor axis of rotation kept it from collecting a notable amount of useful data.

- 1960 – The first weather satellite to be considered a success was TIROS-1, launched by NASA on April 1. TIROS operated for 78 days and proved to be much more successful than Vanguard 2. TIROS paved the way for the Nimbus program, whose technology and findings are the heritage of most of the Earth-observing satellites NASA and NOAA have launched since then.

- 1961 – Edward Lorenz accidentally discovers Chaos theory when working on numerical weather prediction.

- 1962 – Keith Browning and Frank Ludlam publish first detailed study of a *supercell* storm (over Wokingham, UK). Project STORMFURY begins its 10-year project of seeding hurricanes with silver iodide, attempting to weaken the cyclones.

- 1968 – A hurricane database for Atlantic hurricanes is created for NASA by Charlie Newmann and John Hope, named HURDAT.

- 1969 – Saffir–Simpson Hurricane Scale created, used to describe hurricane strength on a category range of 1 to 5. Popularized during Hurricane Gloria of 1985 by media.

– Jacob Bjerknes described ENSO by suggesting that an anomalously warm spot in the eastern Pacific can weaken the east-west temperature difference, causing weakening in the Walker circulation and trade wind flows, which push warm water to the west.

- 1970s Weather radars are becoming more standardized and organized into networks. The number of scanned angles was increased to get a three-dimensional view of the precipitation, which allowed studies of thunderstorms. Experiments with the Doppler effect begin.

- 1970 – NOAA National Oceanic and Atmospheric Administration established. Weather Bureau is renamed the National Weather Service.

- 1971 – Ted Fujita introduces the Fujita scale for rating tornadoes.

- 1974 – AMeDAS network, developed by Japan Meteorological Agency used for gathering regional weather data and verifying forecast performance, begun operation on November 1, the system consists of about 1,300 stations with automatic observation equipment. These stations, of which more than 1,100 are unmanned, are located at an average interval of 17 km throughout Japan.

- 1975 – The first Geostationary Operational Environmental Satellite, GOES, was launched into orbit. Their role and design is to aid in hurricane tracking. Also this year, Vern Dvorak develops a scheme to estimate tropical cyclone intensity from satellite imagery.

- – The first use of a General Circulation Model to study the effects of carbon dioxide doubling. Syukuro Manabe and Richard Wetherald at Princeton University.

- 1980s onwards, networks of weather radars are further expanded in the developed world. Doppler weather radar is becoming gradually more common, adds velocity information.

- 1982 – The first Synoptic Flow experiment is flown around Hurricane Debby to help define the large scale atmospheric winds that steer the storm.

- 1988 – WSR-88D type weather radar implemented in the United States. Weather surveillance radar that uses several modes to detect severe weather conditions.

- 1992 – Computers first used in the United States to draw surface analyses.

- 1997 – The Pacific Decadal Oscillation was named by Steven R. Hare, who noticed it while studying salmon production patterns. Simultaneously the PDO climate pattern was also found by Yuan Zhang.

- 1998 – Improving technology and software finally allows for the digital underlying of satellite imagery, radar imagery, model data, and surface observations improving the quality of United States Surface Analyses.

- – CAMEX3, a NASA experiment run in conjunction with NOAA's Hurricane Field Program collects detailed data sets on Hurricanes Bonnie, Danielle, and Georges.

- 1999 – Hurricane Floyd induces *fright factor* in some coastal States and causes a massive evacuation from coastal zones from northern Florida to the Carolinas. It comes ashore in North Carolina and results in nearly 80 dead and $4.5 billion in damages mostly due to extensive flooding.

21st Century

- 2001 – National Weather Service begins to produce a Unified Surface Analysis, ending duplication of effort at the Tropical Prediction Center, Ocean Prediction Center, Hydromete-orological Prediction Center, as well as the National Weather Service offices in Anchorage, AK and Honolulu, HI.

- 2003 – NOAA hurricane experts issue first experimental Eastern Pacific Hurricane Outlook.

- 2004 – A record number of hurricanes strike Florida in one year, Charley, Frances, Ivan, and Jeanne.

- 2005 – A record 27 named storms occur in the Atlantic. National Hurricane Center runs out of names from its standard list and uses Greek alphabet for the first time.

- 2006 - Weather radar improved by adding common precipitation to it such as freezing rain, rain and snow mixed, and snow for the first time.

- 2007 – The Fujita scale is replaced with the Enhanced Fujita Scale for National Weather Service tornado assessments.

- 2010s - Weather radar dramatically advances with more detailed options.

References

- Fahd, Toufic. : 815. Missing or empty |title= (help); |contribution= ignored (help), in Morelon, Régis; Rashed, Roshdi (1996). Encyclopedia of the History of Arabic Science. 3. Routledge. ISBN 0-415-12410-7.

- Raymond L. Lee; Alistair B. Fraser (2001). The Rainbow Bridge: Rainbows in Art, Myth, and Science. Penn State Press. p. 156. ISBN 978-0-271-01977-2.

- Jacobson, Mark Z. (June 2005). Fundamentals of Atmospheric Modeling (2nd ed.). New York: Cambridge University Press. p. 828. ISBN 978-0-521-54865-6.

- Ronalds, B.F. (2016). Sir Francis Ronalds: Father of the Electric Telegraph. London: Imperial College Press. ISBN 978-1-78326-917-4.

- "History of Meteorological Services in India". India Meteorological Department. March 19, 2016. Archived from the original on March 19, 2016. Retrieved March 19, 2016.

- Dorst, Neal (May 5, 2014). "Subject: J6) What are some important dates in the history of hurricanes and hurricane research?". Tropical Cyclone Frequently Asked Questions:. United States Hurricane Research Division. Archived from the original on March 19, 2016. Retrieved March 19, 2016.

- Nicholas W. Best, Lavoisier's 'Reflections on Phlogiston' II: On the Nature of Heat, Foundations of Chemistry, 2016, 18, 3-13. In this early work, Lavoisier calls it "igneous fluid".

- Anne E. Egger and Anthony Carpi: "Data collection, analysis, and interpretation: Weather and climate". Visionlearning.com (January 2, 2008). Retrieved on 2013-11-06.

Permissions

Index

A

Absorption, 17-18, 23-24, 29-30, 36-37, 40, 45-46, 58, 60, 65, 138

Agricultural Meteorology, 12

Air Mass, 23, 64, 66, 79-84, 87, 98-99, 102, 104, 131, 134, 142-143, 145, 159-160, 162, 166, 171, 200, 205, 208-210

Altitude Variation, 59, 88

Atmosphere of Earth, 14, 87

Atmospheric Chemistry, 1, 14, 52-56, 214

Atmospheric Pressure, 4, 14, 18, 21, 52, 64, 87-91, 159-160, 169, 206, 208, 218

Aviation, 11, 19, 87-88, 93, 112, 124, 181, 196, 202-203

Aviation Meteorology, 11

B

Boiling Point of Water, 91

Boundary Layer Meteorology, 10

C

Changes Due to Global Warming, 110

Changes Due to Urban Heat Island, 110

Circulation, 4-5, 19, 24-25, 30, 33, 35, 46, 51-52, 58-59, 63, 65, 77, 84, 113, 134, 138, 159-161, 166-167, 169, 176, 212, 219, 224-227

Climate, 7, 9-10, 12, 25, 27, 54, 56, 61-73, 77, 90, 94, 98, 103-104, 108-110, 112, 114, 140-142, 146, 148, 150-151, 156, 168, 213, 221, 224-225, 227-228

Cloud, 2-5, 7, 19, 68, 73, 76-77, 81, 83, 86, 92-93, 97, 99-102, 104, 107, 113-125, 127-147, 149, 161, 163, 167-169, 171-173, 176, 179, 185-186, 188-189, 193, 196, 199, 201-206, 211, 213-214, 220, 224, 226

Coalescence and Fragmentation, 143

Cold Front, 83-87, 100, 123, 149, 162, 171, 193, 200, 208-211

Coloration, 121, 125, 136, 138-139

Convection, 32, 45-46, 49, 85-86, 99, 102-103, 108, 116, 119-121, 123, 125, 131, 143, 145, 150, 162-164, 166, 171-172, 207, 211

Cyclogenesis, 113, 159-162, 169-173, 203-204, 210

Cyclone, 4-6, 52, 97, 105, 112, 147, 149-150, 159-171, 174, 176, 179, 183, 193, 195, 199-200, 205-208, 217, 223-224, 227-228

D

Density, 10, 16-17, 20-22, 27-32, 34-36, 38-44, 46, 48, 50, 65, 79-82, 145, 162, 171, 192, 208, 216, 222-223

Depletion, 27, 53, 60-63, 102

Deserts, 69, 80, 104-105, 112, 146, 156-157

Dew Point, 64, 92-99, 113, 134, 143, 163, 210

Diamond Dust, 102

Distribution, 12, 33, 58, 72-73, 105, 111, 133, 135-137, 141, 144-145, 147-148, 158, 163, 177, 194, 223

Distribution in The Stratosphere, 58

Droplet Size Distribution, 141, 144

Dry Line, 83, 85, 210-211

Dust Devil, 167, 176

Dynamic Meteorology, 10, 13, 212

E

Effect on Agriculture, 109, 155

Emission, 7, 24, 31, 34, 36, 43, 77, 99, 107, 143

Environmental Meteorology, 12

Exosphere, 14, 16-17, 20, 27-29, 36

Extremes on Earth, 77

F

Fire Whirl, 168

Formation, 2, 45, 52, 58, 73, 85, 99, 113, 115, 118, 128, 130-131, 133-137, 142, 149, 159-163, 165-166, 169-174, 186, 193, 210, 214-216

Frontal Activity, 102, 145

G

Gauges, 106-107, 142, 151-152

Global Scale, 9

H

Hail, 76, 83, 97-98, 100-101, 103, 115, 126, 139, 144-145, 149, 167, 173, 176, 179-181, 195-196

Hydrometeorology, 1, 12

I

Ice Pellets, 98, 100, 145

Intensity, 31, 43, 58, 60, 71, 78, 100, 103, 107-108, 123, 125, 130, 137, 142, 144-145, 150, 153-154, 162, 168, 175, 191, 227

L

Local Variation, 90

Luminance, 137

M

Maritime Meteorology, 12

Mass, 1, 14-16, 19, 21-23, 28-30, 45, 47-48, 51-52, 54, 64, 66, 72-73, 78-84, 86-87, 90, 92-93, 98-100, 102, 104, 113-115, 125, 131, 133-134, 142-145, 159-160, 162, 166, 171, 200, 205, 208-211

Mean Sea Level Pressure, 87-88, 204

Measurement Based on Depth of Water, 91

Mesocyclone, 167, 169, 172-173

Mesocyclones, 159-160, 162, 167-169, 172

Mesoscale, 9-10, 12, 77, 85, 157, 159, 162, 164, 167, 169, 171-173, 192, 194, 204, 210-211

Mesosphere, 14, 16-20, 29-31, 34-36, 45, 63, 88, 113-114, 117, 137, 139

Microscale, 9, 77, 161, 169, 172-173

Microscale Meteorology, 9, 77

Military Meteorology, 12

Monsoons, 4, 73, 157, 219

N

Nuclear Meteorology, 12

O

Occluded Front, 83-84, 102, 145, 149, 162, 171, 209-210

Orographic Effects, 103

P

Paleoclimatology, 64, 70

Polar Desert, 156

Precipitation, 1, 7, 9, 11-12, 41, 64-65, 67-69, 72-73, 76, 80-81, 83, 85-87, 97-111, 114, 116, 120-123, 125, 127, 129-131, 135, 141-143, 145-160, 173, 176, 179, 181, 187, 191, 194, 196, 201-202, 204-205, 207-208, 210, 223, 227-228

Pressure, 1, 4-5, 7, 10-11, 14, 16, 18, 20-23, 28-29, 34, 46-52, 57, 64, 72-73, 76, 78, 82-85, 87-93, 95, 97, 102, 105, 115, 120-123, 134-135, 145, 147, 157, 159-163, 165-167, 169-176, 187, 192-193, 199-212, 218-219, 222, 224

Principal Layers, 16, 19

R

Rain, 2-3, 5, 7-8, 11-12, 53, 67, 69, 73, 76-77, 83, 87, 97-98, 100, 102-107, 109-111, 113-115, 118, 123-126, 138-139, 142-147, 149-159, 165-166, 173-175, 179, 182, 187-188, 193, 195, 214-217, 219, 225, 228

Raindrop Impacts, 145

Raindrops, 100, 144-145

Rainforests, 105, 147, 157

Recreation, 182

Reflectivity, 137-138, 153

Refractive Index, 24

Remote Sensing, 8, 152

Renewable Energy, 12

S

Scattering, 23, 42-43, 102, 139

Snow, 2-3, 8, 11-12, 71, 73, 76, 80-81, 97-98, 101-102, 104, 106-107, 112, 115, 123, 125-126, 145, 149, 174-175, 178, 180-182, 187, 195, 207, 216, 218, 225, 228

Snowflakes, 100-102

Solar System, 43, 72-73, 77-78, 114, 141-142

Species, 34, 38, 47, 60, 69, 113-114, 126-131, 136, 202

Speed of Sound, 20-22

Squall Line, 85-86, 131, 200, 210-211

Standard Atmospheric, 87, 90-91

Stationary Front, 83, 85, 170, 209-210

Steam Devil, 168

Storm, 6, 34, 40, 72, 78, 89-90, 101, 103, 108, 113, 139, 150, 154-155, 157, 160, 162-168, 171-179, 182-186, 188-190, 199, 204-205, 210, 222, 226-227

Stratosphere, 9, 14, 16-21, 27, 30, 34-36, 44-47, 50, 56-61, 72-73, 113-114, 117, 134, 136-137, 139, 141, 191, 224

Subtropical, 67-68, 80, 94, 105, 109, 135, 151, 157, 159-160, 162-166

Surface Weather Analysis, 82-83, 170, 187, 199, 204, 207

Synoptic Scale, 8-9, 51-52, 77, 83, 159, 161-163, 169, 204, 207, 210

T

Temperature, 1, 3-4, 7, 10-12, 14-22, 24-25, 28-34, 36, 38, 45-50, 56, 64-65, 67-73, 76-77, 79-81, 83-86, 89-100, 102, 104, 109, 111, 113-114, 134, 142-144, 148-150, 159-160, 163, 165-166, 168, 170-171, 175-177, 180, 182, 187, 192, 195, 197, 199, 201-204, 206, 208-212, 216, 218-219, 221-224, 226

Thermosphere, 14, 16-21, 27, 29-36

Thickness, 19, 21, 37, 47, 50, 56, 58, 85, 138, 200, 206, 211

Timeline of Meteorology, 214

Tornado, 131, 134, 160, 162, 167-169, 173, 176-177, 183, 195, 225, 228

Tornadoes, 8, 73, 86, 88, 126, 149, 159, 167, 169, 172-173, 176, 217, 227

Tropical, 4, 6, 9-10, 46, 50-51, 59, 66-68, 72-73, 79-80, 82-83, 88, 90, 94, 102, 105, 107, 109-110, 112, 120, 135, 145, 147, 150-151, 155, 157, 159-169, 172, 176, 178-179, 181, 185, 192-193, 195, 198-199, 203-204, 207, 210, 212, 217, 223, 225, 227-228

Tropical Cyclones, 68, 88, 102, 105, 109, 145, 147, 150-151, 157, 159-163, 165-167, 169, 172, 176, 178-179, 181, 192, 198, 203-204, 207, 212

Troposphere, 9, 14, 16, 18-21, 25, 30-32, 35-36, 44-50, 59, 72-73, 89, 113-114, 117-119, 127, 133, 137, 139-140, 162, 164, 166, 170-172, 191, 203-204

U

Ultraviolet Light, 56-57, 60

V

Varieties, 106, 114, 128-129, 136, 151, 175, 207

Virga, 99, 107, 117, 121, 130-131, 135, 141, 143, 145, 159

W

Warm Front, 83-85, 87, 100, 104, 121, 123, 125, 162, 171, 193, 200, 208-210

Water-saturated Air, 142

Waterspout, 168, 173

Waterspouts, 160, 162, 169, 173

Weather, 1-2, 4-12, 18-19, 21-22, 46-47, 64-66, 68, 70-73, 76-90, 96-97, 102, 107, 110-112, 114-115, 118, 120-121, 124, 126, 134-135, 140, 142-143, 145, 149, 152-155, 159-161, 163-164, 166, 169-170, 173, 177, 180-181, 183, 185-214, 216-217, 219-228

Weather Forecasting, 1-2, 5-8, 10, 66, 72, 76, 115, 187-189, 191, 193, 195-199, 201, 203, 205, 207, 209, 211, 213-214, 216-217, 225

Weather Front, 64, 80-81, 85

Weather Map, 77, 82-84, 187, 189, 198-206, 209-210

www.ingramcontent.com/pod-product-compliance
Lightning Source LLC
Jackson TN
JSHW052211130125
77033JS00004B/229